BENEATH HILL 60
WILL DAVIES

BANTAM BOOKS

LONDON • TORONTO • SYDNEY • AUCKLAND • JOHANNESBURG

TRANSWORLD PUBLISHERS
61-63 Uxbridge Road, London W5 5SA
A Random House Group Company
www.rbooks.co.uk

BENEATH HILL 60
A BANTAM BOOK: 9780857500496

Originally published in Australia in 2010 by Vintage,
a division of Random House Australia Pty Ltd
First publication in Great Britain
Bantam edition published 2010

Addresses for Random House Group Ltd companies outside the UK
can be found at: www.randomhouse.co.uk
The Random House Group Ltd Reg. No. 954009

The Random House Group Limited supports The Forest Stewardship Council (FSC),
the leading international forest certification organisation. All our titles that are printed
on Greenpeace approved FSC certified paper carry the FSC logo. Our paper
procurement policy can be found at www.rbooks.co.uk/environment

Typeset in Garamond by Midland Typesetters, Australia.
Printed in the UK by CPI Cox & Wyman, Reading, RG1 8EX.

2 4 6 8 10 9 7 5 3 1

Will Davies is an historian and film-maker. He edited the war diaries of E P F Lynch into the acclaimed and bestselling memoir *Somme Mud*, and wrote its companion volume, *In the Footsteps of Private Lynch*. He lives in Sydney.

Acclaim for Somme Mud:

'As haunting and graphic a description of trench warfare as any I have read . . . this is a warrior's tale . . . a great read and a moving eye-witness account of a living hell from which few emerged unscathed' *Daily Express*

'Such is the force of Lynch's direct, compelling account of war . . . we grow to care about him and his companions, and to see what they see' *Guardian*

'In its honesty and earthiness it has quite justifiably been compared to *All Quiet on the Western Front*' *Good Book Guide*

'The voice of an ordinary, but highly literate, private soldier, who simply endured the horrors that surrounded him and got on with his job . . . it truly is "a time capsule"' *Birmingham Post*

'This gripping memoir . . . will have the hairs standing up on the back of your neck . . . [an] excellent book' *BBC who do you think you are?* magazine

'Here is the stink and stench of war . . . horrifying, scarifying and very humbling as well' *Herald Sun*

'Brilliantly evokes the terror, horror, elation, friendship, gore and depression that made a combat infantryman's life so dangerous, so traumatic and, if he survived, so memorable' *Courier Mail*

www.rbooks.co.uk

Also by Will Davies

SOMME MUD
IN THE FOOTSTEPS OF PRIVATE LYNCH

and published by Bantam Books

For my beautiful sister, Bron

Dedicated to the officers and men of the Australian Tunnelling Companies, AIF, who gave their lives in the service of their country and Empire in the Great War. And to Captain Oliver Woodward and the men who returned to Australia wounded, diseased and traumatised, unable to forget the fear and horrors of their war underground.

'Tunnelling was just like a game of chess; one had to anticipate the opponent's move. You didn't always know that you were going to get away with it. All the tension all the time – the strain underground and the darkness. It was terrible. It was not war, it was murder.'

Lieutenant W. J. McBride, 1st Australian Tunnelling Company

CONTENTS

INTRODUCTION

In the earliest days of the First World War, when stalemate set in, a terrible underground war of mining and countermining erupted in tunnels beneath the trench lines. It continued until June 1917, when Captain Oliver Woodward of the 1st Australian Tunnelling Company pushed down a plunger and fired two of 19 massive mines under the German lines at the opening of the Battle of Messines. The largest man-made explosion in history up until that time smashed open the German frontline and enabled the Allies to begin an offensive that would contribute greatly to the final victory at the end of the following year.

This book came about when I received a call from an old mate, David Roach. He had just heard that the feature film *Beneath Hill 60: The Silent War*, for which he wrote the script, was going to go into production. It told the story of the Australian tunnellers at Hill 60, focusing on Captain Woodward, and David wondered if I would write the book on this little-known and yet fascinating aspect of Australia's First World War history.

It is a frightening story of men in tiny tunnels not much bigger than the dimensions of a coffin, 30 metres underground, with water seeping from the cold earth and Germans tunnelling nearby, looking for an opportunity to obliterate them. I hate tight places and am seriously claustrophobic, but it is history that I know and love, and it is an extraordinary story that until now has gone untold. I felt it was important to tell it to the current generation, filling in the details that a feature-length movie cannot, and exploring the real story of the tunnellers.

Research was crucial to this book, and the research lay with just one man: Ross Thomas, the leading authority on the Australian First World War tunnelling companies. While working in Queensland as the inspector of mines in Charters Towers in the late 1980s, he learnt of diaries kept by Captain Oliver Woodward, the man who led the Australian tunnellers at Hill 60 and who had attended the Charters Towers School of Mines before the war. Intrigued, Ross tracked down Barbara Woodward, Captain Woodward's daughter, who provided him with all five volumes of her father's war journals, which were begun in the 1930s. From the start Ross was fascinated with the story revealed in the journals, and he spent the next 20 years learning all he could and collecting and researching the history of the three Australian tunnelling companies. The war journals became the inspiration for the film *Beneath Hill 60*, produced by Bill Leimbach and directed by Jeremy Sims, of which Ross is the executive producer.

Ross very generously provided me with all his research, his contacts and access to his library and Woodward's journals. Soon I was absorbed in a less-appreciated aspect of First World War history. It became a labour of love, much like my previous work editing Private Edward Lynch's account of his time on the Western Front, *Somme Mud*, and writing my follow-up book, *In the Footsteps of Private Lynch*.

This is the story about a small but strategically important

section of the Western Front that was contested by the Allies and the Germans from October 1914, just weeks after the declaration of war. At the time, Australia had hardly moved onto a war footing, yet here, along the Messines Ridge near Ypres, the fighting was already fierce and would remain so until mid-1918.

The focus is the work of the Australian tunnellers under Captain Woodward's leadership, as for eight months they prepared to explode two of 19 massive mines that would blow apart the ridge along nine kilometres of frontline around the Belgian village of Messines on 7 June 1917. As many as 10,000 Germans died in the combined blast, which was heard across England and as far away as Dublin. By the end of that day, the British had advanced in places nearly five kilometres and were claiming it as their best and most successful day up until that time in the war. The Battle of Messines opened the front and ushered in the next great battle, the Third Battle of Ypres – or, as it is more commonly known, simply Passchendaele.

To understand this action, we also need to look at the broader, fascinating history of the tunnelling companies and the stories of the men who traded their mining jobs for a terrible and terrifying silent war primarily fought not in the trenches but in claustro-phobic and treacherous tunnels beneath the frontline.

This, then, is their story.

Prologue

It is just after 12.30 am on 7 June 1917, and Captain Oliver Woodward of the 1st Australian Tunnelling Company is settling into his firing dugout and anxiously beginning his final checks. He has almost three hours to wait, but already he can feel the sweat trickling down the length of his spine and disappearing into the waistband of his corduroy breeches. Although it is mid-summer, a cold chill of fear racks his body.

He looks about him. The dugout has been well prepared for this moment. A small electric bulb casts a yellow light across the other men's faces, which, like his, are tense and frozen in concentration. Two candles add to the meagre glow, wisps of grey smoke spiralling up and disappearing into the gloom of the low ceiling. The two officers from his unit plus a British brigadier general are all motionless, waiting, their breathing imperceptible, listening, thinking and focused. In the dark corners of the dugout, outside the ring of light, are some of the men of his unit. They sit silent and expectant, their job as tunnellers now done.

Oliver Woodward's mind is focused and clear – clearer than he can remember for some time. He thinks through the details of the two massive explosive charges he has the responsibility to detonate: in the Hill 60 chamber, 45,700 pounds of ammonal and 7500 pounds of guncotton. In the other, the Caterpillar mine, another 70,000 pounds of ammonal. Both galleries have been stacked carefully – he ensured that, just as he supervised the installation of the long electrical leads and fuse back to this firing point, and of the detonators with their three fail-safe back-ups. And he had overseen the testing – yes, the endless testing.

All looks good, but as Woodward rehearses in his mind the firing procedure, still he wonders if it will all go off.

He does not have long to wait.

Spread across the front, 100,000 men lie on the start line for the attack, stretching south to take in the Messines Ridge. Another 115,000 are there to provide support. They, too, are sweating and fearful of the terrible day of fighting they know is ahead, and they, too, wonder about their chances.

At 1.15 am, Oliver Woodward again tests the resistance in the electrical leads. He needs an assurance that the circuit going from his dimly lit dugout – down, down deep into the ground, along a damp, muddy tunnel, under tightly packed sandbags and on to the detonator in the underground gallery packed with explosives – is ready for firing. So he tests to make sure that the very weak circuit is still flowing through it and tickling the detonator – not enough to fire it, just enough to show that the circuit is intact. He trusts no one, not even himself. All he trusts is the equipment he holds in his muddy hands, which indicates that the leads are sound, the current is getting through and the circuit is complete.

His other worry is the Germans. The listeners have heard them. He has heard them. Their relentless digging. The nearly imperceptible bite of their tools into the grey-blue clay, the rattle of their winch and the quiet bump of their bucket as it lands at the bottom of their shaft. They are edging closer. It will be touch and go. He withdrew his listeners only at the last minute so the men could complete laying the sandbags, but according to his calculations, the Germans are about six feet from the Allies' explosive-laden mine system.

Years of work have led up to this moment, and many lives have been lost: the lives of tunnellers, good men – his men – in the dark, dank suffocating tunnels below this tortured hill. Their work is done. It is now up to him.

At 2.10 am, an hour before zero hour, the troops guarding the mine entrances are withdrawn, counted out and sent to their posts.

2.25 am. Woodward hopes the leads stretching under the German frontline are still intact. With 45 minutes left, he completes the last resistance check. It is now in the hands of chance and, of course, the Germans.

Finally, he carefully attaches the leads to a small hand-held igniter.

Outside all is ominously quiet apart from the odd Allied shell falling far behind the lines, searching out German fatigue parties struggling up to the front. All Woodward can hear is his heart pounding – pounding in his chest and reverberating in his ears – and the watch, ticking in a different rhythm, as though in competition with his heart. The watch will decide the moment for the devastating blast.

The watch ticks on. Tick . . . tick . . . TICK . . . TICK.

'Five minutes to go.' The authoritative voice of Brigadier General Lambert breaks the silence. Woodward is in his firing position. All is ready.

'Three minutes to go.'

Woodward looks sideways at the two officers to his right, an exploder at their feet in case the electrical charge fails.

'One minute to go,' calls Lambert.

Woodward's hand moves into position on the exploder, his fingers slowly gripping the cold brass handle. He feels the sweat on his palms as he tightens his grip. The silence is deafening.

'Forty-five seconds.'

'Twenty seconds.'

'Ten seconds, nine, eight, seven, six, five, four, three, two, one – *fire!*'

Down goes the firing switch. At first, nothing. Then from deep down there comes a low rumble, and it is as if the world is being split apart.

ONE

Mobility to Stalemate

Unlike the many inexperienced teenagers who rushed to enlist at the outbreak of the First World War, 28-year-old Oliver Woodward was already out in the world, making his mark. His skills as a mine manager had been recognised by the Mount Morgan Gold Mining Company, and when they opened new copper mines in Papua in January 1914 they chose him to supervise operations at a number of them and to work as the assistant geologist. It was a difficult and isolated life for Woodward at Laloki, about 37 kilometres to the southeast of Port Moresby. He oversaw hundreds of Papuan miners, all the while coping with hot, humid and uncomfortably tropical weather. His commitment and diligence made him invaluable, and he was quickly given more responsibility.

On 3 August 1914, the Australian steamer *Matunga* arrived at Port Moresby, and Woodward left on horseback early the following morning, accompanied by a few Papuan staff, to collect supplies. On reaching the port, he was handed a message from the head office of the Mount Morgan Gold Mining Company, telling him: 'In view of the European War Crisis you will discontinue all operations immediately, realise all assets and return at first opportunity.'

'This indeed was a bombshell,' Woodward wrote in his diary. 'On hearing no rumours of War . . . I concluded that I was probably the first of the civilians in Port Moresby to know of the outbreak of War.'[1]

Germany was now suddenly the enemy, and German New Guinea lay just to the north. Woodward had to act quickly. Many of the 200 indentured Papuan mine workers had travelled from the Fly River district hundreds of miles to the west to take up jobs at the mines, so first he assured their welfare by securing government jobs for them. He quickly disposed of the company assets and by the end of the week was ready to leave. He headed to Port Moresby to catch the *Matunga*, which was still in the dock, back to Australia.

Fearful that the German fleet would arrive at any moment, the colonial administrators organised a home defence unit. All able-bodied men were called to defend the town, and Oliver Woodward happily volunteered – on the proviso that should the opportunity arise for him to return to Queensland, he would be honourably discharged. The home defence force placed the letters AC, for Armed Constabulary, on their shoulders and headed off to guard the radio station and patrol the beaches. But with only one ancient and totally useless cannon, they stood little chance of success should the might of the German fleet arrive.

'Night duty in the tropics was rather pleasant and as we patrolled Ela Beach, when the stillness of the night was broken only by the gentle lapping of a coral-fringed sea on a beautiful white sandy shore, it seemed a wild stretch of imagination to picture ourselves as defenders of the outpost of our Empire,' wrote Woodward.[2]

He borrowed a whaleboat from the owner of the Mount Diamond Mine so that he could have one of the Papuan mine staff – referred to in his diary as Head Boy Paul – transport his personal effects from Laloki to Port Moresby. He had instructed Paul to leave for Port Moresby after midnight so that by dawn the boat would be around the headland and could make use of the early breeze. Paul dutifully agreed and set off into the night. Woodward was off duty that night, and when the Ela Beach patrol saw the large boat moving slowly towards the shore, they thought this was the anticipated German invasion. Fortunately, before the firing started, the boat was recognised, but Oliver Woodward received a stern reprimand from the harbour master for his failure to obtain the necessary clearance.

As Port Moresby was a naval fuelling depot, the unexpected arrival on the horizon of a ship always created great excitement in the town. So when two warships were spotted in late August, less than a month after the declaration of war, an increased sense of fear and apprehension gripped the populace. There was relief when it became clear that they were the Royal Australian Navy ships HMAS *Australia* and HMAS *Encounter*, which were on their way to Rabaul to capture the German garrison and the radio station. The ships would be able to provide protection for SS *Matunga* to leave Port Moresby, where she had been stuck ever since the declaration of war. The *Matunga* would sail for Queensland the following day (17 August 1914). Upon hearing the news, Woodward quickly obtained leave from the Armed Constabulary and secured a berth on the already heavily booked passenger ship.

The *Matunga* sailed from Port Moresby through the night to Cooktown. She made it through without coming under German attack, but two years later would not be so lucky. In August 1916, returning from Rabaul she was captured by the German raider

Wolf. The *Matunga* was forced to go westward to Waigeo Island, off the remote northwestern tip of West Irian, where the crew, the passengers and the cargo were offloaded and taken aboard the *Wolf.* The Germans sank the *Matunga*, and the *Wolf* returned to Kiel in February 1918 to a hero's welcome. She had made the longest sea voyage of a warship during the First World War and had returned with not only substantial quantities of rubber, copper, zinc, brass, silk, copra, cocoa and other essential materials for the German war effort, but also 467 prisoners of war. All of her crew were awarded the Iron Cross and her commander, Captain Nerger, the highest German decoration, the *Pour le Mérite.* The captured crew and passengers of the *Matunga* remained in Germany for the duration of the war.

Woodward's journey on the *Matunga* lasted two days: he disembarked at Cooktown and made his way to Mount Morgan, a booming mining community south of Rockhampton, to report to his bosses. He spent a fortnight debriefing them about the Papuan mining operations, and then, sick with continual bouts of malaria, which he had contracted while away, he headed south to visit his family in Tenterfield, in northeastern New South Wales. There he rested and began his recovery.

He had grown up in rural Tenterfield, and in 1901, at the age of 15, was sent to board at Newington College, a Methodist school for boys in the Sydney suburb of Stanmore. It was a boarding school based on the classic English model, with little comfort: boys were given the bare minimum of food; they froze in winter and sweltered in summer; discipline was harsh, and humiliation was considered character building. Boarders learnt fast that to survive you needed independence, strength of body and character, rat cunning, and a certain degree of invisibility. Woodward excelled as an athlete, was awarded colours in rugby and cricket, and served in the cadets. He received a trophy for shooting, an important skill for young men at this time. A boy from the bush, he was self-assured and confident, and could handle a rifle and a horse.

The post-Victorian era was a time of morality, Imperial splendour and social change. Woodward was brought up to be loyal to Britain and a royalist, yet to be a proud Australian. Though at Newington he would have rubbed shoulders with the old money of Sydney – the sons of the squattocracy and the wealthy Protestant immigrants – he retained the egalitarian nature of his rural roots.

After school he had worked in mines at Irvinebank, North Queensland. Then he headed north to Charters Towers, in Queensland. There he worked underground as a labourer for three years, mining gold and studying at the Charters Towers School of Mines part time to become a mining engineer.[3] He received high marks and was awarded two prestigious medals, the W. H. Browne medal for mining in 1909 and the medal for metallurgy in 1910. After further experience underground, he qualified as a mine manager and worked in copper mines at Mount Morgan and Broken Hill before being sent to the newly opened mines in Papua.

Now back in Australia, Oliver Woodward was faced with a dilemma. Copper was an important ingredient for munitions, so his job was a protected occupation.

On the one hand, he believed this was 'a very slender excuse for a young and able-bodied man'. Yet he 'honestly felt that there was some reasonable argument' why he should at least delay enlisting: 'The general opinion was that the Australians, due to lack of training, would never be active participants in the fighting but that garrison duty in Egypt would be their role. I felt that I was not justified in sacrificing my professional career merely to seek adventure.'[4] So he returned to Mount Morgan and his work underground.

It was not long before he was receiving white feathers anonymously in the mail.

✕

At first it was expected that the war would be over by Christmas. But by November, the British and French armies found themselves mired in a stalemate with the Germans.

Flurries of snow were making life terrible; men froze and their feet and lower limbs went black with frostbite and trench foot. Neither side was making progress, but there were a few active hot spots along the front. One of the hottest was a small area hardly bigger than a football field: the notorious and deadly Hill 60.

Height is everything in war. It is important strategically, whether it be a small platoon-sized engagement or a full-scale battle. Holding the high ground not only allows you to fire *down* on your enemy while they have to attack *up*, but it gives you the advantage of being able to observe your surroundings. The higher ground south of the town of Ypres and the village of Zillebeke, in Belgium, was intersected by a railway line that ran from Ypres to Comines on the France–Belgium border. A cutting had been made through it, and the spoil had been deposited on each side of the railway cutting, forming slightly higher hillocks. The hillock to the east of the railway line was 60 metres above sea level: it became known as Hill 60. The spoil dump on the western side of the line, which was S-shaped, was dubbed the Caterpillar. It was 55 metres above sea level. (These man-made high points were similar to the famous Mound at St Eloi, just to the north along the same ridgeline, or the much fought over

Butte De Warlencourt between Pozières and Bapaume, on the Somme.)

Hill 60 was crucial high ground as it provided excellent views to the north across the British lines at Zillebeke, and of Ypres, four kilometres away. Ypres was of paramount importance for it stood between the German army and the English Channel. The Allies simply could not afford to lose Ypres.

To the southwest of Hill 60 and the Caterpillar was the Messines Ridge, stretching between the villages of St Eloi and Messines. It was only of modest height – at Messines village it was 60 metres above sea level. What made it so important in the First World War was its height relative to the flat lands around it.

When Britain had declared war on Germany, it had not expected to make any real military contribution, certainly not in Europe. Other European nations, especially France and Germany, had military strategies in place, but Britain had relied instead on accords and agreements with its European allies. It also relied heavily on the power of the navy should a show of strength be required. With only a small army focused on colonial problems – most recently the Boer War in South Africa – Britain was unprepared for a large-scale European war. The hastily formed British Expeditionary Force (BEF) consisted of only one cavalry and six infantry divisions, a total of about 86,000 men. Initially it was anticipated they would be absorbed into the French army, holding the extreme left of the French line in northwestern Belgium through to the coast, where it was believed little fighting would take place.

The Germans, however, had something else in mind. They had been developing since 1900 the Schlieffen Plan, in which Germany would invade France by first crossing Belgium, thus evading massive concrete underground defensive systems constructed by France in the early 1870s, then wheel southwest and pass to the west of Paris to cut off the French army. The Germans felt they needed to knock France out of the war quickly, as the Russian army was a threat in the east.

By 20 August, Brussels had fallen to the Germans. Then, with an eye on the Channel ports, they pushed westward. On 23 August 1914, the BEF made contact with the rapidly advancing German army at Mons, but were pushed back in 'the great retreat' to virtually the outskirts of Paris. Here the BEF and the French held the Germans at the River Marne before launching a counterattack, pushing the Germans back 50 kilometres to the high ground on the other side of the River Aisne. As the British and French armies struggled to take the German high ground, the frontline stalled and the armies dug in. This was the end of the German advance and their chance for a quick knockout blow against France. Suddenly Germany found itself committed to a war on two fronts that would not, as hoped, be over by Christmas. And the beginnings of trench warfare brought with it stalemate.

Two massive Russian armies had pushed westward into Prussia, but between 26 and 31 August 1914, the Germans had encircled the Russian 2nd Army at Tannenberg, capturing 100,000 men. Fearing further annihilation, the Russian 1st Army then fell back into East Prussia and was defeated at the Battle of Masurian Lakes, with the loss of a further 125,000 men and 200 guns. This quickly ended the threat of Russian occupation, taking some of the pressure off the Germans on the Eastern Front.

The only real movement on the frontline now was westward. The Germans, under General Falkenhayn, were trying to outflank the Allies in 'the race for the sea'. Their aim was to take the

Channel ports of Dunkirk and Calais, enabling German ships to operate in the English Channel, hence cutting off supplies to Britain, France and Belgium. Ypres, which was held by the Allies, stood directly in the Germans' path and was the key to this part of the front. On 17 October, General Falkenhayn launched a massive offensive supported by heavy artillery and eight divisions. Thrown into the battle were poorly trained German troops, many of them young university students. They failed to take the British line that extended in an arc around the city. This was the beginning of what became known by the Allies as the First Battle of Ypres but by the Germans as *Kindermord von Ypres*, 'the massacre of the innocents at Ypres'.

Next, seven German divisions attacked on a narrow front between Messines and Gheluvelt. Their objective was the high ground that stretched from the village of Passchendaele, northeast of Ypres, to the southern end of the ridge, at the town of Messines. The Germans attacked westward on 21 October, quickly taking some small villages to the east of Messines, but their attack up the ridge was halted by British troops dug in along a line of disconnected trenches.

Determined to take the ridge, the Germans brought up heavy artillery and pounded the shallow British trenches from close range. Then wave after wave of fresh German troops assaulted the thin line of defenders, and after savage hand-to-hand fighting the British fell back, abandoning Messines and retiring to the nearby village of Wulvergem. To the north at Wytschaete, Indian troops also put up a gallant fight, but they too were driven off the ridge, and the high ground fell to the Germans.

By early November, the lines had stabilised, with the Germans holding the ridgeline extending north from Messines through Wytschaete and on to Hooge, while the British retained the high ground to the southwest at Hill 63, near Ploegsteert Wood, which enabled them to observe their front along the Douve River. The French held Hill 60.

By 22 November 1914, the First Battle of Ypres was over, and the British held a fragile line around the town. The Germans, however, now held the high ground extending along more than 20 kilometres of the line, and behind this, some of the most important and productive industrial regions of France and Belgium. High Command on both sides saw Ypres as crucial to their future strategy, but it had become a symbol of British defiance and would remain so for the rest of the war.

After just four months, the frontline ran more than 650 kilometres, from the Swiss border in the south to the coast between Dunkirk and Ostend. Trench lines, breastworks and a mass of supportive infrastructure were starting to evolve, but for the most part, the location of the line would not change for the rest of the war. Both sides had fought to a standstill and accepted the new military state of stalemate.

✕

Participating in the First Battle of Ypres was a German private named Adolf Hitler, who was a member of the List Regiment. The regiment had a brief period of training, then after being reviewed by the Bavarian King, Ludwig III, and his son Rupprecht, the commander-in-chief of the German 6th Army, they marched 40 kilometres to the frontline near the villages of Besalaere and Gheluvelt, to the north of the Menin Road.

Here they were immediately thrown into the battle and suffered appalling casualties. They lost more than a third of their troops, 349 men, on the first day of the attack. More than half of the regiment's officers were killed, including the popular Colonel

List. By the time Hitler's shattered regiment was relieved on 1 November and marched back to Wervick, between Menin and Messines, more than two-thirds of its men were dead, wounded or captured, and it could no longer be considered a functional frontline unit. Hitler rose from infantryman to dispatch runner, was promoted in the field to the rank of corporal (which he remained until the end of the war) and was nominated for the Iron Cross Second Class. But most of all, against all odds, he survived.

After a week's rest, the List Regiment, under a new commanding officer, took over trenches at Messines on 8 November and then positions near Wytschaete opposite the French. It was here that Hitler won his Iron Cross, supposedly for placing himself in front of his commanding officer to protect him from machine-gun fire. Undertaking the dangerous job of delivering messages between his rear headquarters and the frontline, Hitler survived a number of narrow escapes, each one reinforcing his belief in divine providence and that he was being preserved for greater things later in life.

Though brave and resolute, Hitler was not popular. He harangued his comrades about their lack of commitment and bravery, their need to confront danger and death, and the general conduct of the war. These rantings not only left him friendless but probably adversely affected any chance of promotion. He spent his spare time reading newspapers and books on politics, and painting watercolour scenes around this sector of the front. One of these, of the destroyed Messines church, is today in a museum in the town.

Finally, on 20 November, after 18 days in the frontline, the shattered List Regiment was relieved and moved to rest areas in the rear. Here they received new reinforcements, and the men recovered from their wounds, undertook training and prepared for the next move to the front. This was not long in coming and Corporal Hitler was soon back running messages in the frontline and defying death.

Though the frontline was largely static, smaller-scale offensives were carried out on occasion in certain strategic places. On 10 December 1914, the Germans captured both Hill 60 and the Caterpillar from the French, and the Messines Ridge. The British, extending their line towards Ypres, took over from the French and immediately began making plans to wrest Hill 60 and the Caterpillar back from the Germans. Included in these plans was the extension of some shallow defensive tunnels begun by the French in the opening stage of the war. In the months and years to come, these small hills would become some of the most tunnelled and mined areas of the Western Front. Oliver Woodward's skills as a mining engineer would soon be in demand on the battlefield.

Two

The War Goes Underground

Ironically, the man who perhaps did the most to initiate a British response to the German tunnelling threat on the Western Front came not from Allied High Command. He was in fact a large, brusque and domineering Conservative MP named John Griffiths, later known as John Norton-Griffiths.

Griffiths was a man of the age: strong, educated, flamboyant and an adventurer. Strongly built and over six feet tall, with dark hair and clear green eyes, he was 43 years old when war broke out. In 1888, at the age of 17, he had sailed to South Africa and worked in remote parts of the country as a miner and in engineering and excavation. He fought in Africa in the Second Matabele War of 1896 and later in the Boer War. He had a great sense of Empire and Britain's part in the world, and soon found himself building railways and other engineering projects in Africa and America. This led him to found his own company, Griffiths and Co., a structural engineering firm specialising in major infrastructure projects, including tunnelling.

Since the early 1900s, 'Empire Jack', as Griffiths was affectionately known, had been doing very well. He had contracts all over

Britain – including the construction of the City of South London Underground, the Battersea Power Station and the sewerage system for Manchester – and in other parts of the world. A project to build railway and port facilities in Australia was abandoned when war was looming.

In the days leading up to the declaration of war, Griffiths advertised in the London press for the raising of a private irregular regiment made up of veterans who had served in the various recent South African wars. Men flooded in, and hordes of British and colonial veterans pushed and shoved at his recruiting tables to be selected. The 2nd King Edward's Horse was formed. Griffiths, who'd had military training in South Africa, enlisted in his own regiment and gave himself the rank of major. He put £40,000 into equipping the regiment and became the second-in-command with Lieutenant Colonel Montagu Craddock the commanding officer.[1] Griffiths was soon to move on to another wartime role, though.

Griffiths had men working on the Manchester sewer system and the thought came to him that these skilled tunnellers might also be able to make a contribution to Britain's war effort. Tunnelling in Manchester and for the London Underground involved working in clay, which posed problems in tight and low spaces where there was no room to swing a pick. Clay tunnellers had developed a unique system of digging known as 'clay kicking' or 'working on the cross'. Lying on his back at 45 degrees, supported by a T-shaped backrest, a tunneller would use his feet to dig the clay with a special lightweight pointed spade. The spade had a longer handle and a footrest on which the tunneller placed his booted foot and then pushed, kicking the spade away from his body. The clay fell away as he dug and was collected, bagged up and taken back along the tunnel using light rail with small rolling stock. The men pushed the tunnel forward, picking up the backrest and repositioning it as they progressed.

In mid-December 1914, Griffiths wrote to the War Office requesting permission to take a handful of 'moles' – as clay kickers called themselves – to France to try the clay-kicking technique in a frontline situation. However, Field Marshal Sir John French, the commander-in-chief of the BEF, found little of interest in the suggestion and filed the letter away.

The use of tunnellers and engineers in the military had a long history, but for decades their role had been neglected and the British army had put few resources into training or equipping them.

The task of early military tunnellers was to build underground concealment for weapons, food and ammunition, headquarters, billets and even hospitals. Military mining developed in response to the construction of castles and fortified towns that, if self-sufficient in water and food, could hold out against attackers virtually indefinitely, resulting in sieges often lasting years. Military engineers built catapults, siege towers and missile launchers to try to break down defensive walls, but what proved most effective was tunnelling underneath them.

The introduction of gunpowder, in the mid-13th century in England, and cannons rendered tunnelling obsolete. But as fortifications were strengthened and military engineers looked at new designs and new ways to counter the effects of the cannon, the idea of tunnelling beneath defensive works was revisited from the mid-16th to mid-17th centuries, especially during the English Civil War.

In Britain, the first military engineers, the forerunners to the

Corps of Royal Engineers, were appointed by William the Conqueror some time after 1066. They did not hold military rank and were outside the permanent standing army – basically they were civilians who were called upon to do specialised tasks for the military as required. In the 1760s, when things became dangerous for the civilian tunnellers on the Rock of Gibraltar, they downed tools and left. The army realised they needed men who had discipline and were subject to military control, and so in 1772, the first Company of Soldier Artificers was raised, made up of carpenters, blacksmiths, stonemasons and miners.

By 1813, this unit was renamed the Corps of Royal Sappers and Miners (the word 'sapper' came from the Italian 'zappa' or spade). Sappers had the dangerous job of digging shallow trenches towards enemy fortifications from a trench line. But just as soon as tunnelling units were formed, along came the explosive shell, which made even the toughest castles and fortifications vulnerable and eliminated the need to tunnel under and destroy them from below. Siege warfare became a thing of the past, and mobile armies with artillery, cavalry and troops meant that there was little chance for stalemate or time to tunnel. Training fell off and these units reverted to other specialist engineering tasks such as building fortifications, bridges and roads, their tunnelling skills forgotten. It is perhaps not surprising then that when the first troops of the British Expeditionary Force sailed for France in August 1914, the sapper units had only very limited training in mining, and only under safe and ideal conditions.

✕

At the end of 1914, the entrenched armies along the Western Front were settling in for an uncomfortable festive season. Snow lay in the hollows and was blown by gusty winds from the north.

South of Ypres, near the village of Festubert, in France, units of the Indian Corps from the hot, dry Uttar Pradesh region were suffering in rudimentary trenches that offered little protection from the biting wind, let alone the Germans a few hundred metres away.

It was here that the first Allied mining operation of the war was about to be attempted. In conjunction with a planned attack, the Allies dug a shallow tunnel running out about 20 metres, an estimated four metres from the Germans' trenches. A small charge of explosives – 20 kilograms of guncotton – was placed into position in readiness for the attack, but an enemy mortar bombardment forced the abandonment of the firing position and the mine. It was never fired – an unfortunate start to Allied mining operations in the Great War.

The Germans had also seen the merit in attacking from underground. On 20 December, east of the village of Le Plantin, very close to the site of the failed British effort at mining, three mysterious coloured flares lit the night sky, soaring high and hanging there, bright and mesmerising. Suddenly, nearly a kilometre of Indian trench erupted as a series of ten charges exploded along the frontline. The blasts had a devastating effect on the troops, the Germans estimating that 3000 Indian soldiers died, many still sitting in the trenches, apparently suffocated.[2]

Not a man to be brushed aside, Griffiths had again written to General French, pressing the advantages of his specialist miners. Events now overtook him, with the explosion of the German mines under the Indian Corps at Festubert, but General French was at his headquarters at St Omer, and Griffiths' letter took a week to be opened and read.

French, of course, was under pressure to act from his immediate staff, who in turn were under pressure from everyone down the line, through to the men in the frontline trenches. If the cold weather, regular artillery strafing by the Germans and the frightful conditions in the trenches were not enough, now they were forced to live with the fear of being swallowed suddenly by a massive explosion beneath their feet. After the mine explosions at Festubert, the Indian Corps refused to remain at the front and withdrew their forces into reserve – a serious step that GHQ did not wish to see repeated. French knew something must be done to address the threat, but he was a man of the old school, and rather than take up Griffiths' offer he passed the responsibility of tunnelling on to the ill-equipped Royal Engineers.

Due to inexperience and lack of skills and equipment, the Royal Engineers were already struggling to cope with the demands of constructing defensive works and drainage systems, and of maintaining roads and troop accommodation. Offensive mining was outside their skill set, and they did not have enough men.

But, orders being orders, the Royal Engineers applied themselves to the task. In early January 1915 at Rue du Bois, near Armentières, the 20th Fortress Company started work on a tunnel towards the German frontline. As hard and fast as they dug, however, the water poured in, flooding the tunnel. Their antique pumps could not keep up, and soon the morale and energy of the men faltered and the work slowed. Morale took a further blow when the Germans hoisted a sign, written in English: 'No good your mining. It can't be done. We've tried.'[3]

Various brigade commanders began forming their own mining sections, made up where possible of men from mining areas around the United Kingdom, but their lack of training and equipment, plus the appalling winter conditions, rendered their attempts unsuccessful.

And then, six weeks after the Festubert explosions, the Germans did it again, exploding mines under English regiments at St Eloi on the Messines Ridge on 3 February 1915. A week later they exploded a second mine near the same point in the line. Both caused severe English casualties.

Griffiths was at home and going crazy. He was obsessed about getting his moles to France. As his wife said, 'He was nearly like a maniac, frantic for action.'[4]

In mid-February, a telegram arrived that summoned him to the War Office for a meeting with his old friend Field Marshal Lord Kitchener. They had met in South Africa during the Boer War, when Kitchener was commander-in-chief, and in early 1914 in Egypt when Kitchener, then the consul-general in Cairo, invited him to discuss plans for the Aswan Dam.

Sitting in plush red leather chairs in Kitchener's large office, the two men discussed the perilous situation in France. The weary Kitchener shoved a sheaf of papers into Griffiths' hand: communiqués from General French at his headquarters in St Omer outlining the German mining successes. Then, bending forward, Kitchener spoke in a low voice and asked for advice, something he rarely seemed to do. Griffiths' answer was immediate and to the point: 'The only thing I suggest, sir, is that we use moles,' he said, to which Kitchener replied, 'What on earth are moles?'[5]

Grabbing the coal shovel from the grate, Griffiths threw himself onto the floor and proceeded to demonstrate the art of clay kicking to an amazed Kitchener. Usually a man of great caution and slow reaction, Kitchener was immediately won over, clearly understanding the possibilities of the clay kickers and their potential. 'Get me ten thousand of these men,' he said. 'Immediately.'[6]

Within hours, Griffiths was on his way to France, and the following day he walked into the St Omer office of the engineer in chief, George Fowke, and his assistant, Colonel Robert Harvey. They were all ears. Again Griffiths demonstrated the clay-kicking action, explaining the ability of his moles to dig narrow tunnels very quickly and attack the Germans unexpectedly, thereby taking back the initiative. Not only this, but his men could break into German tunnels and destroy their underground workings.

The following morning, 14 February 1915, Griffiths, along with his foreman, Mr Miles, and his assistant engineer, Mr Leeming, was taken to four frontline engineering headquarters to demonstrate the technique, finally stopping at Givenchy, not far from where the Indian troops had been blown up the previous December. Keen to understand the soil and the conditions for digging, his foreman, Mr Miles, dropped to the ground, collected a sample and let it spill from his hands. 'It's ideal, isn't it, Miles?' Griffiths said, to which Miles smiled and replied, 'It makes my mouth water.'[7]

Things began to move fast: the War Office approved the formation of a specialist tunnelling unit as an offshoot of the Royal Engineers, and Leeming returned to London to close down the sewer contract and dismiss the men, but then immediately offer them a posting to the front.

Griffiths was beside himself with excitement, as he could now apply his energy to the problem with the knowledge that he had full support at all levels of command. Back in England, he found 20 volunteers waiting for him. Eighteen of them were quickly issued their kit and rifle and dispatched to France. On Thursday 18 February 1915, these men had been safely digging sewers beneath Manchester. Five days later they were tunnelling towards the German lines at Givenchy.[8]

The glorious history of the tunnellers had begun.

THREE

In the Darkness and Mud

The War Office sanctioned the formation of eight mining companies, each made up of six officers and 227 men. These men were coal, slate and tin miners, all strongly unionised and politically active. And they were not the usual malleable young recruits of 18 or 20. They were men in their 60s, grey-haired and toothless – they had trouble eating the hard frontline rations – who had little idea about military etiquette and cared even less about it. They had an independent attitude and would not be bullied. Realising this, the War Office became 'frightened by the need to accept a sudden influx of untrained, fiercely free-thinking miners' who, when brought together without military training, could become 'undisciplined mobs'.[1] So the War Office quickly demanded that recruiters choose carefully and allow only reliable and experienced men to enlist. The commander-in-chief 'was known to be specially against having Scotsmen'.[2] There were not enough clay kickers available in Britain to make up even one company, so Griffiths raced up and down the Western Front in a muddy Rolls-Royce, calling upon men with underground work experience to transfer to his new companies, and men were formed into units as quickly as possible.

By early April 1915, the British were almost ready to launch an ambitious mining operation beneath Hill 60, intended as the start of the battle that would put Hill 60 back into the Allies' hands. Clay kickers in one of the new mining units, the 171st Tunnelling Company, had dug shafts down to about four metres and were now digging tunnels towards the German lines. The middle tunnel, known as M1, went straight under the hill. The other two tunnels, M2 and M3, ran to positions along the German front-line. Each of the three main tunnels had two tunnels branching off from it, and five mines were to be laid.

In the days before the attack, the Germans could be heard working very close to M3 – so close that the clay kickers could hear their whispered conversations. It was thought they would break through into the British gallery at any time. Speed was now a matter of life and death, so the British tunnellers ramped up the pace. Instead of the average progress of 1.8 metres a day, they achieved 3 metres, 3.6 metres, 4.2 metres and once even managed 4.8 metres in one day – an extraordinary effort. But each day, the Germans, too, advanced their tunnel. The British and the Germans were now separated by just a few metres of sticky clay.

Two miners, sappers Morgan and Garfield, were inching the M3 tunnel forward when suddenly they broke through the clay, leaving a yawning black hole in the face of the tunnel. Immediately it dawned on them: they had broken through into the Germans' gallery. They quickly put out their candle and lay silently in the dark, dank confines of the tunnel, their eyes and ears straining for any light, any sound. Sure enough, they heard the heavy tread of boots then the splash of water as a German advanced through the tunnel towards them. Would he see the fractured tunnel wall – the gaping hole that now joined the two workings – and come to investigate? Morgan and Garfield knew they had to get out, and fast.

They ran back along the British tunnel as quietly and quickly as they could. Up the shaft and out into the air they raced, and into the officers' dugout, where they reported to Second Lieutenant Thomas Black. He said there was no real option but to go back down and take the fight to the Germans. With their hearts booming in their chests, they each took a pistol and headed back down the shaft. This time they could not risk a candle or a torch, so they moved forward quietly into darkness, easing themselves along the wet, seeping tunnel, trying to keep their weapons clear of the mud.

When sapper Morgan felt they had gone far enough and would be near the hole, he suggested Black turn on his torch and investigate. No sooner had the light touched the close-pressing walls than a shot rang out. The bullet ripped through Black's uniform and embedded itself in the wall. Off went the torch, and the three men scrambled back along the tunnel as fast as they could – little concerned about their obvious and noisy flight – up the shaft and into the trench and daylight.

There they thought through their options. Realising that the Germans would be as scared of them as they were of the Germans, Black led the two reluctant tunnellers back into the darkness, back along the long tunnel and back to the face. The German had gone – but in his place he had left an ominous 'camouflet', a type of explosive charge that caused damage to an enemy's tunnel without breaking through to the surface. They quickly cut the wires leading back to the German line, to neutralise the charge. Then they laid a larger explosive and played out their own leads back to a dugout, and waited until the attack to fire it.

Luck was with the British tunnellers. At 7 pm on 17 April, the mines were successfully fired and the British assault troops raced across no-man's-land. The mines at M1 caused a massive hole to be blown in the side of Hill 60. Those German defenders who did not die in the explosion were dazed and quickly driven from the position. There were few British casualties.

Hill 60 was back in Allied hands . . . but for how long?

There could be no assurances in this part of the front, for in the two months leading up to the offensive, the British and Germans had been neck and neck with their mining operations and attempts to shift the line. When the first of the clay kickers – 33 men of the newly formed 171st Tunnelling Company – had arrived at Hill 60, a British Royal Engineers officer, Lieutenant White, had already begun expanding the maze of shallow tunnels taken over from the French in late 1914. Attached to the 28th Division, White had pushed a tunnel forward and under the German frontline, which he blew on 17 February 1915. It had done little damage, but it was the first successful British mine blown to that point in the war.[3] On the same day the Germans quickly struck back, firing a mine beneath the British frontline salient at Zwarteleen, a few hundred metres from Hill 60. An assault party took the crater, but the British drove them off.

The clay kickers had been rushed to Hill 60, arriving on Monday 22 February 1915. As if by way of welcome, on that day the Germans blew the largest mine in the war to date, at Shrewsbury Forest, just two kilometres to the east of Hill 60. It inflicted severe casualties on the 16th Lancers and demolished a long section of their trench. The Germans then rushed and occupied the crater, driving back British counterattacks and forming a new frontline.[4]

On 4 March, the 171st Tunnelling Company had had the distinction of blowing their first mine beneath Hill 60. The relatively small charge – 60 kilograms – achieved its purpose of significantly damaging the German underground workings and making it difficult for them to mine.

Immediately, the British started work on another tunnel, which stretched 36 metres, to within three metres of the German frontline. The construction of this longer tunnel showed up problems that would plague the miners well into 1917: bad air and poor ventilation. But here the earth was made up of hard, dry sand that was so firm it did not need to be shored up with timber reinforcing, and this allowed the miners to push the tunnel forward at a rapid pace.

Just 72 hours later, the mine was ready, the men withdrawn and the infantry prepared. The enemy was heard tunnelling beneath and to one side of the tunnel, so the attack was brought forward. At 7.40 pm on 7 March 1915 an electrical charge ignited 87 kilograms of gunpowder, forming a crater in the German line 14 metres in diameter and wiping out a considerable proportion of the enemy trench. This was the first mining attack to damage German trenches and the first to be followed by a planned infantry attack.

During March, British mining activity increased at Hill 60 and along the front, at St Eloi and Ploegsteert on the Ypres salient and at Givenchy, Houplines, Fauquissart and Cuinchy a little further south.[5]

At St Eloi on 14 March, the Germans hit back against the heavily defended British position on 'the Mound'. It was much like Hill 60: a modest slope perhaps only ten metres above the surrounding area, but strategically important. The Germans followed up with a successful assault, capturing not only the Mound but also one entrance to a British tunnel. Losses were heavy, with more than 500 men killed, wounded or missing. The British counterattacked, but the Mound stayed in German hands.

Soon after, the Germans fired another mine at Zwarteleen, the village bordering Hill 60. Their surveying was poor, and the mine did as much damage to the German trenches as to the British, but it affected British morale and the men became jumpy.

And this pattern – one step forward, two steps back – continued after the successful mining operation at Hill 60 on 17 April. Within a short time, the British success took on the smell of a British blunder. Though the hill had been recaptured, it now protruded into the German frontline, and like all dangerous salients, it was directly in the line of concentrated enfilading fire from both flanks. No sooner had the shattered Germans recovered from the explosion of the mines than they began a relentless fire into the exposed new British positions, raking the trenches and parapets with shot and shell. As the firestorm ended, the British lined their trenches, huddled below the shattered sandbags that ran along their parapets, to await the anticipated German counterattack.

The Germans, in well-rehearsed style, pushed back against a British line bristling with rifles and machine guns. Between 17 and 21 April there was a relentless to-and-fro battle for this churned patch of mud, scooped out by mining and sown with the bodies of the dead. Charles Bean, Australia's official First World War historian, would later describe it as 'a rubbish heap' in which it was 'impossible to dig without disturbing a body'.[6]

On 22 April, a heavy bombardment of Ypres by the Germans with 17-inch shells heralded the start of the Second Battle of Ypres. The French still retained responsibility for the front in a section of the line to the northeast of Ypres, near St Juliaan, and troops from the colony of Algeria had recently arrived there. Late in the afternoon, as men shuffled into position ready for the dusk 'stand to', there appeared a yellow-green mist moving slowly and low to the ground. Initially thought to be a smokescreen for the German infantry attack, on the spring breeze it drifted into the frontline trenches.

For a moment the Algerians smelt pepper – no, it was pineapple – and felt a burning sensation in their throats and pain in their chest. Quickly this strange, nauseating smoke overtook them and they threw down their weapons, leapt from their trenches and ran screaming to the rear. They were the first victims of a new and frightening weapon: chlorine gas.

A six-kilometre gap suddenly appeared in the Allied line. Even the Germans were surprised by the success of their gas experiment. If they had actually exploited the break in the Allied line, they could have created a serious breach and potentially a catastrophic collapse along the whole front. Instead, the newly arrived Canadian Division quickly replaced the French colonials and, although also gassed, they held the line. Some of the gas also drifted southwest and was blown across the fatigued British troops on Hill 60. By this time, it had lost its potency, but the experience was enough to give them a sense of the dangers to come.

On 4 May, almost two weeks later, the exhausted men holding the crest of Hill 60 were woken by a sentry who noticed an ominous cloud drifting up from the German lines. 'Gas! Gas!' he shouted. As the men rushed to the parapet, they could clearly hear the audible hiss of the gas as it left its canisters. Quickly they started to choke and vomit as the gas mixed with water in their lungs and converted to hydrochloric acid, stinging and burning their throat, chest and lungs. Light exposure to the gas would cause vomiting and irritate their eyes; moderate exposure would cause damage to the lungs and possibly pneumonia later; and a few deep breaths of a high dose would be lethal.

There had been no gas masks when the first chlorine gas attack occurred in April 1915. At first, temporary pads were fashioned on the battlefield, the earliest often soaked in urine to neutralise the acidic chlorine gas. Some men used socks, handkerchiefs and flannel belts, which they soaked in a solution of bicarbonate of soda and tied across their faces when the gas gong was sounded or gas was seen approaching the trenches. It was difficult to fight with such rudimentary masks, so effective gas masks and anti-asphyxiation respirators were developed by the British and were on general issue by July 1915. The Germans then developed phosgene gas and mustard gas, which caused damage even to soldiers wearing gas masks as they burnt not only the throat and lungs but also any exposed skin.

At Hill 60 in May 1915 the cost to the British was great: 90 died and a further 58 succumbed to gas poisoning at a nearby dressing station. Hundreds more were retching and vomiting a strange green slime as they clawed at their throats and gasped for air. Others wandered off to the rear, struggling alone in search of stretcher-bearers or a Casualty Clearing Station, but there was little, if anything, the medical staff could do.

Within 15 minutes, Hill 60 was back in German hands.

FOUR

The Experience of Gallipoli

While the British had already learnt from circumstances in France and Belgium that they needed to take the war underground, in April 1915 the Australians were just beginning their own painful lessons. After training in Cairo and landing on Gallipoli, the Australians quickly became bogged down. The Turks contained the Allied forces to the steep slopes above the beach, and stalemate, much like that on the Western Front, set in.

In the days after the landing, AIF field companies – specialist engineer units – dug artillery positions, trenches, the beginnings of two roads and wells, organised beach defences and started the first dugouts into the hillsides that were to prove so important in the months ahead. Because of the danger of snipers on the high ground above the beach, the Allied trench system quickly had to increase in sophistication. The frontline was pushed forward by building 'saps': trenches dug towards the enemy, from which lateral trenches branch off to form a new frontline. Once forward, a new parapet could be built. 'Looped' firing positions were constructed: a small hole between sandbags, or steel plates with holes through which snipers could fire their rifles with protection.

As the tunnels moved forward, it was found that there was no need for timber supports or revetting as the earth was dense alluvial compacted sand, similar to sandstone, which was very stable yet easy to dig. This was fortunate as there was very little available timber on the peninsula. Although the tunnels did not require supports, progress was slow, with only a metre or so being completed over a 24-hour period.

Initially, the idea of offensive mining was not even considered. Then reports came in of sounds of digging beneath Quinn's Post. New Zealand engineers dug three shafts and established listening posts. The Australians of the 15th and 16th Battalions who were stationed at Quinn's Post included many miners.

These big, staunch fellows, though genial and often humorous, were hard-grained men accustomed to form their own opinions and not afraid to express them. In spite of the confidence of the authorities, some of these men took their own precautions. So it was that on May 17th, a man of the 15th, by the name of Slack . . . distinctly heard the steady, persistent, muffled knocking of the enemy's picks. Slack summoned his CSM, Williams, and CO, Sampson. Both heard the sound and it was duly reported.[1]

The Australians dug a tunnel towards the Turkish positions and blew a camouflet. The next day, further Turkish activity was heard. This time, members of the New Zealand Field Company fired a camouflet.

The Turks, keen to drive the invaders back into the sea, had realised the value of mining. Unlike the Australians, though, they did not have a mining industry from which to recruit experienced men. They seemed not to worry about being heard, and the 13th Battalion history notes that even during a thunderstorm they 'could hear the Turks digging under Quinn's'.[2] The history goes on to say: 'We counter-sapped for all we were worth to meet the Turk underground and at 2.30 pm on the 28th May, we blew his sap in.'[3]

Their work did not stop the Turkish mining. On the following day, the Turks exploded a mine beneath Quinn's Post, and all the Australians in this part of the line were killed or wounded. The Turks followed up with an assault but, after savage fighting, the Australians beat back the attack. Thirteen Australians were killed and 81 wounded, and Turkish casualties were estimated to be more than 300. This was the beginning of the offensive tunnelling that would dominate Allied and Turkish operations until the end of the year.

After a number of futile frontal attacks against well-sited Turkish positions, the British had realised that they only had two options if they were to break out: open a new front and go around the flanks of the Turks, something they would try at Suvla Bay, or dig under and crack the Turkish line by breaking through from beneath. Within a month of landing, the AIF field companies had begun to focus on tunnelling, and this would become their most important function until the evacuation in December 1915.

The Anzac Corps commander, Lieutenant General Birdwood, ordered that men with mining experience be released from their battalions to form specialist mining units.[4] These units were sent to the frontline at Quinn's, Courtney and Pope's Posts to begin offensive tunnelling. Many of the men had worked as miners in places such as Kalgoorlie, Broken Hill, Cobar, far north Queensland around Charters Towers, and the goldfields to the northwest of Cairns. This gave them a clear advantage over their Turkish enemy. 'As far as defensive and offensive mining was concerned, the supremacy of the Anzac miners was almost complete,' wrote

Charles Bean. 'Whenever the enemy were heard tunnelling, they simply waited till the sound came from a few feet, and then blew in his workings.'[5]

Bean's description of Slack, the man who raised the alarm over Turkish digging under Quinn's Post, paints a picture of the kind of men who formed these early mining units. A 45-year-old railway ganger, Corporal Slack 'was a tall, sinewy fellow, older than most, with the humorous, kindly wrinkles of a typical miner and had constantly refused promotion in the AIF. He had mined all over Australia, and of late had managed a small Tasmanian tin "show".'[6]

The Australians invented a system of driving underground tunnels forward instead of open saps. Transverse tunnels extended out to each side, very close to the surface. When an attack was to be launched, the roofs of these tunnels were removed, leaving an open trench ready for use. This meant assaults could be launched much closer to the Turkish trenches, minimising the time men spent charging across no-man's-land. This became so successful that the commander of the 1st Australian Division ordered that tunnelling rather than sapping was to be the accepted practice all along the front.

On 24 June 1915, the Australians fired their first offensive mine, and a further five mines were blown in the days after.[7] 'The Anzac miners began to have a marked effect upon the Turkish tunnellers who were reported to be in a state of semi-panic,' wrote Bean.[8] In late July the Turks blew a mine only 12 metres from their own trench, sending their barbed wire flying into the air. A couple of days later, Australian engineers digging towards Lone Pine waited until 'the Turks came within five feet of them, and then fired a countermine, the Turkish picks being heard until the moment of the explosion'.[9]

The Turks were aware of the aggressive scale of the Allies' tunnelling. It is believed they had been informed by a spy in Egypt who had probably heard about the tunnelling from the wounded returned to hospital there.[10] In response, Turkish Command had issued an order to begin defensive mining in all areas where the frontlines were close. At Lone Pine, the Turks were not aware of the exact location of the tunnels, but they did notice that a great deal of digging was in progress. A Turkish commander commented after the war: 'Every day when I looked at the English lines and saw always mound upon mound of new earth rising, I said to myself, "What can these English be doing? I shall certainly wake someday and find that they have tunnelled to Constantinople."'[11]

It was decided to land a British force to the north of Anzac Cove at Suvla Bay and for the Australians to undertake a major feint attack along their front. Apart from some small gains, this proved unsuccessful, as did the landing at Suvla, and so from August there was little heavy fighting above ground.[12]

✕

The stalemated situation emphasised the need for Australia to develop its own specialist tunnelling units, similar to Griffiths' clay kickers who had formed companies attached to the Royal Engineers in France and Belgium.

The idea was championed by two men, Tannatt William Edgeworth David, and Ernest Willington Skeats. Before the war, Sydney University professor Edgeworth David was already well known as an academic, an Antarctic explorer and a geologist. He had come to Sydney from Wales in 1882 to take up an appointment for the state government as the Assistant Geological Surveyor. In 1907–08, he accompanied Ernest Shackleton and

Douglas Mawson to the Antarctic, where he led an expedition up Mount Erebus and attempted to reach the South Pole. By the outbreak of the war, Edgeworth David was the head of the university's geology department; he was influential, highly regarded around the world and well connected politically, involving himself in recruiting rallies and the raising of patriotic funds.

Joining the army as a major in October 1915, Edgeworth David was the geologist attached to the Tunnelling Battalion and left for Europe on the same ship as Woodward in February 1916. His job as a military geologist was to give advice on the location of ground water and the siting and design of trenches and tunnels.

Ernest Skeats was the head of the geology department at Melbourne University. In August 1915, he and David suggested to the Minister of Defence, Senator George Pearce, that a specialist mining and tunnelling unit be formed. More formal than the ad hoc mining companies already formed at Gallipoli, it would be independent of the army and made up of Australian miners. In September 1915, the prime minister offered Britain 1000 expert miners and mining engineers, plus equipment, for service in the Dardanelles or elsewhere. The British government gladly accepted the offer, asking that the men be provided in the form of tunnelling units.

Within a week, more than 50 miners had volunteered at Broken Hill, and others were standing by awaiting details of enlistment. The Defence Department decided to establish a mining battalion headquartered in Victoria, comprising three companies: one made up of miners from New South Wales, the second of miners from Victoria and South Australia, and the third miners from Queensland and Western Australia. This was a different structure to that adopted by the English and Canadians, who had set up numerous tunnelling companies that were independent of each other and not part of an overall battalion.

The AIF's upper age limit, at that time 38 years, was extended to 50 years.[13] To be commissioned, prospective officers needed to be experienced mining engineers or managers, or to be mining surveyors with experience underground. Non-commissioned officers and sappers were to be men 'experienced in underground work as foremen, shift bosses, face men, tunnellers, carpenters and blacksmiths'.[14] Such men already serving in the AIF were allowed to transfer into the new battalion. Australia quickly began seeking men with vision and experience in mining to form the three tunnelling companies.

Oliver Woodward's moment had come.

FIVE

White Feathers
and the
Call to Arms

In the year since he had returned to Australia to work at the mine at Mount Morgan, Oliver Woodward had continued to feel conflicted about his role in the war. Though copper was desperately needed for the war effort and his contribution to mining was more crucial than if he had become a foot soldier in the first days of the war, it played on his mind that all around him men were enlisting to serve overseas.

It was Gallipoli that changed everything for Woodward. Suddenly, Australian men were giving their lives in their thousands, showing enormous courage in appalling and treacherous conditions. The AIF's fighting men had proven themselves, and it was now clear that Australia's expected role of 'garrison duty in Egypt' had been replaced by a far heavier burden.

'Following the glorious landing of the Anzacs at Gallipoli and the publication of the first casualty list, it seemed that the time had arrived when selfish personal interests should be cast aside and when service to one's Country became of prime importance,' Woodward wrote.[1] But the real turning point came when he received a telegram on 21 August 1915 to say that his cousin,

Major Moffat Reid, had died on Gallipoli. 'This sad news turned the scale in my desire to enlist and without delay I told the General Manager of the company, Mr. Boyd, my intention.'[2] Boyd was sympathetic, as his son had lost an arm at Gallipoli, and so Woodward's employment was terminated on 9 September.

A week later, Woodward left for Tenterfield to spend time with his family. Then he took the train to Sydney and did four weeks' training at a rifle club. These clubs were not run by the army but were certainly encouraged by them as they prepared men for the military by giving them basic training such as drill with weapons and musketry practice. In October, he signed up at Victoria Barracks. 'Until I finally enlisted I was never happy or contented,' he wrote.

> [It] probably caused a wave of patriotic satisfaction in the breasts of the Mount Morgan senders of White Feathers . . .
>
> In regard to these White Feathers I have one wish and one regret. The wish that the senders may some day in their life experience a fraction of the agony which their thoughtlessness brought to the recipients, and the regret that I did not have the feathers to carry with me as a pillow.[3]

With his extensive mining experience, he reported to the officer in charge of the training school for engineer officers of the new Australian Mining Battalion. And here Woodward had the first of many experiences that made him critical of the army's training and traditions. He wrote: '[I had] a list of qualifications and experience in mining which would permit me filling, with reasonable prospect of success, almost any mining job in civil life.'[4] Yet he was informed that being a hard rock miner, it was doubtful he would be useful to the new Mining Battalion, which would likely be working in sand, soft soil and clay. 'This was the first of many occasions when I was forced to admire the success of the Military Organisation in placing square pegs in round holes,' he reflected.[5]

Fortunately, his experience in managing large numbers of men was recognised and he was sent to the Officer Training School for the Mining Battalion. Woodward and a group of men with similar experience were given the non-commissioned rank of second corporal, just one rank up from private. Along with a number of second lieutenants they were marched from Victoria Barracks to the nearby Sydney Cricket Ground and placed in one large dormitory under the Members Stand. The Members Dining Room became the Officers Mess. Here Woodward first encountered the military pecking order. The second lieutenants were not happy sharing their mess with mere second corporals. 'For a time the atmosphere was tense,' he wrote, 'but we 2nd Corporals found satisfaction in the realisation that it was easy to create 2nd Lieutenants but difficult to create gentlemen.'[6]

Overseas, the Australians had become renowned for being anti-authoritarian and defying the rules. The training camps were no different, as Woodward learnt when he and the other second corporals were given guard duty. Men who were on leave were due back at 11 pm, but many would arrive 'in the wee small hours of the morning', carrying their leave pass in one hand and a packet of fish and chips in the other. 'Being true soldiers, and knowing in full the advantage of taking the initiative, the Fish and Chips would be forced into one's hand while the Sapper streaked off to his tent.'[7]

Woodward found the army's training seriously lacking. In addition to the usual marching and basic training, he found himself learning skills that he felt little prepared him for an engineer's war, let alone that of an officer, such as the tying of ropes, knots and lashings. Once he reached the muddy trenches and dark tunnels of the Western Front he would remember with a degree of amusement the hours he spent learning how to construct wooden observation towers.

In particular, he was deeply concerned about the preparedness of the officers who would be responsible for large numbers of men on the field of battle:

> The weakness in our training lay in the fact that from the Colonel
> downwards, all officers were amateurs, each knowing about as
> much as the other in regard to Military matters. Without any
> reservation, I definitely say that so far as training was concerned,
> no single officer of the Mining Battalion was qualified to leave
> Australia with commissioned rank.[8]

Yet on the morning of 23 December 1915, he and 20 other young officers were promoted to second lieutenant.

> I was appalled by the thought that I was about to embark for
> Active Service as an officer whose knowledge of Military Training
> was so lamentably poor. Can you wonder that I experienced a
> feeling of doubt as to my right to take charge of a body of men in
> actual War when I knew so little of Military matters? The fact that
> the majority of officers and men of the Battalion made good is no
> answer to the arguments that undue risk was taken. Even to this
> day the grounds of this appointment remain a mystery.[9]

✕

For Oliver Woodward, the arrival of the New Year was filled with sadness. He was in Tenterfield on his last leave before he would depart for the war. Having just had Christmas dinner with his family, he was unsure when, or if, he would ever see them again. He wrote many years later that the memory of his anguish at parting with his family was still etched in his mind and would remain with him forever.

No doubt also on his mind was the girl he loved, Marjorie, in north Queensland. The two had known each other for many years but only recently had their relationship blossomed. Marjorie's father, William Waddell, had met the Woodward family decades before and had even nursed Oliver when he was a baby. Waddell was a Scotsman who had moved with his three brothers to join their cousin John Moffat, who managed a mine at Irvinebank in north Queensland.

It was not until June 1907, when Woodward was 21, that he met William Waddell's daughter, Marjorie, for the first time. They were guests at a 'Grand Picnic' at John Moffat's property, Loudoun House, to mark the official opening of the Stannary Hills to Irvinebank Tramway. Woodward, who was mining in Irvinebank and attended the party, recalled: 'The younger folk romped and danced in the dining room . . . Among those whom I danced with was a nine year old, auburn haired girl Marjorie. On this our first meeting, I found her to be a charming child, mature beyond her years.'[10] Little did he imagine that this sweet child would grow up to be the woman he would marry.

After Woodward had finished his studies at the School of Mines in Charters Towers, he was a frequent visitor to Loudoun House, and it was here that he met Marjorie again in late 1911, when he was 26 and she 14. 'In the interval of time, Marjorie had developed and was now a charming girl. I enjoyed her company so much that I had begun to regret the disparity in our ages,' he wrote.[11]

Moffat resigned his position as manager of the Irvinebank mine and was replaced by Woodward's uncle, John Reid. Woodward was sent to Koorboora, where the Waddells lived, to redraw plans for a tungsten mine. It was decided he should stay at the Waddells' as the local pub was of a poor standard. Woodward wrote:

> I was delighted with the thought of being in the house with
> Mr. and Mrs. Waddell as I had much regard for the family and
> in particular for Marjorie. There was something stronger than
> friendship for this lovely girl in her teens. When my thoughts went
> beyond the bonds of friendship, I came up against the disparity in
> age which in those days seemed a blank, impenetrable wall.[12]

The task of redrawing the plans was completed within a week, the time passing 'pleasantly, but all too quickly' for Woodward.[13]

When he returned to Australia after his time in Papua New Guinea, Woodward moved to Broken Hill to gain experience in underground mining. In February 1915, he travelled to Sydney and stayed with relatives at Cremorne Point. Here he was surprised to learn that Marjorie Waddell was in Sydney attending Miss Hale's Business College. It had been three years since he had last seen her. 'In that period of time, the memory of this charming girl frequently came to mind,' he wrote. 'I confess my heart missed a beat when I held her hand in greeting.'[14] He was relieved she had 'no inkling' of his regard for her 'other than that of friendship'. Between February and September 1915, Woodward spent eight days in Sydney and met Marjorie as often as possible.

Woodward wrote to Mr and Mrs Waddell, asking their permission to write to Marjorie. So it was on 11 May 1915 that he wrote his first letter to her, beginning a correspondence that would continue right up until their marriage in 1920.

When Woodward was at the Rifle Club, the pair regularly met in the grounds of Government House to share sandwiches in the gardens. In due course Marjorie finished her classes and had to return to her family in north Queensland. Woodward saw her off on a ship. 'After the ship sailed, I set off for Moore Park depot feeling very depressed,' he recalled.[15]

Now, as the New Year dawned and they were just beginning to get close, war was about to separate them.

On 2 January 1916, Second Lieutenant Woodward returned from Tenterfield to Victoria Barracks. Next day, he took the train to Liverpool and reported to the massive army camp at Casula. On these hot, dusty paddocks the other ranks had been undergoing their training. Men from across Australia had poured in, most without any military training, many unable to read or write, but with mining skills and experience in working in claustrophobic conditions underground. Now the men and the officers who would lead them were to begin their joint training. The battalion comprised 51 officers and 1075 other ranks. The following morning, Woodward was allocated to 1st Australian Tunnelling Company, which comprised four sections. He joined No. 1 section.

Initially, training focused on marching and drill, which Woodward, like all soldiers, quickly came to loathe: 'The monotony of this soon became almost unbearable. As a 2nd Lieutenant my position was at the rear of the Section and I had to patiently march behind, turning, wheeling and halting at the word of command.'[16] Then commenced a period of training in military mining. He was introduced to the Wombat Borer, a motor-driven auger with a diameter of about 20 centimetres that was used to dig a narrow borehole from an Australian trench to the enemy's. A torpedo-like explosive charge was pushed into this hole, then fired remotely from the Allied line.

The 'Wombat', which had been developed in secrecy by Captain Stanley Hunter of the Geological Survey of Victoria, had been used with limited success at Gallipoli. The hope now was that it could be rapidly deployed on the Western Front as an offensive weapon. But as it could drill no more than about 60 metres, it could only be used where the opposing frontlines were close together. And operating the machine close to a hostile enemy position was likely to be problematic, especially given the amount of noise and smoke produced by the motor, which was crude by today's standards.

Second Lieutenant Woodward and other men with mining experience were not impressed by the machine. Woodward tried to imagine what a German sentry, gazing through his periscope across no-man's-land, might make when he saw a drill head, deflected upwards, burst through the surface and leap two metres into the air, thrashing and gyrating. Would he think this was some new Allied secret weapon when in reality it was just the good old Aussie Wombat Borer? Only experienced miners were chosen to run the Wombat Borer, yet in training aspects of its operation were considered unsafe. Even before the men set sail for the war, the future for the Wombat Borer looked decidedly shaky. Nevertheless, training in its use took up much of Woodward and his men's time.

The military ranking system continued to grate on Woodward. As a second lieutenant, the most junior rank for a commissioned officer, he was 'not admitted to the charmed circle, the members of which were Colonels, Majors and Captains'.[17] He went on to note wryly that when it came to paying mess dues, though, the second lieutenants and senior officers had equality.

Many of the men saw being in the army as like any other job, with officers as bosses. After providing their eight hours of labour a day, they felt they should be free to do as they pleased. Trying to impose military discipline onto these free-spirited and independent men was difficult. One night, Woodward gave instructions to be woken in time to take charge of a 4 am shift overseeing a training exercise. He was duly awakened, but by the persistent shaking of the tent flap and a very Australian voice bellowing, 'Eh, mate, are you awake?' A private soldier calling an officer 'mate' sounded sacrilegious to Woodward's ears. But later, in the mud of the Somme or the deep tunnels beneath Hill 60, he would come to know that 'no officer could ask for a higher compliment than to be considered a mate of his men'.[18]

Woodward himself sometimes found his superiors' demands for formality ridiculous. While overseeing men using the Wombat Borer one day, he issued the command, 'Push her in.' This brought a severe reprimand from his commanding officer. 'Advance the drill' is what he should have said. 'I cannot put in writing the terms used by the Sappers!' joked Woodward in his diary.[19]

Some of the men at Casula had such little regard for authority that they even mounted a strike, although Woodward noted in his diary that his men refused to join it, which brought them high praise from the public and the press. 'We were hailed as Soldiers true and staunch,' wrote Woodward.[20] The reasons they stayed at their posts were, however, perhaps less than heroic. He believed that they resented the fact that other men had organised the strike without first consulting them and that 'they were heartily sick of Camp life and feared that participation in the strike would postpone the date of embarkation'. He quipped: 'That a miner would purposely miss participating in a strike shows what he thought of Camp Life!'[21] So for Oliver Woodward and the men of the Mining Battalion, there was a sense of relief when it was announced that a date had been set for their embarkation: 20 February 1916.

SIX

The
Earthquake Idea

At Gallipoli, after the Allies' failure to break through at Suvla in August 1915, tunnels had been started from a number of additional positions along the Australian front, including the Nek, Lone Pine and Johnston's Jolly. Near Lone Pine, ammonal – a compound of ammonium nitrate, aluminium and TNT that the British had recently begun using on the Western Front – failed to blow through to the Turkish trenches above, and when a number of men went into the tunnel, they died after breathing in carbon monoxide gas, which was given off by exploding mines and camouflets. In late November, the Australians blew a charge of 225 kilograms of ammonal at Russell's Top, which ended Turkish tunnelling in this area. The Australians also laid explosives in a number of large, deep galleries along the Turkish front. These mines were never fired, but had they been blown simultaneously the effect would have been similar to the huge explosions at Hill 60 and all along Messines Ridge in June 1917.

With the frontlines still stalemated and the Australians taking terrible casualties, preparations began for the evacuation of Gallipoli, and the mines took on a new role: to be ready should the Turks attack during the evacuation. Rather than offensive tunnelling, the miners and engineers turned their attention to building better accommodation for the diggers, whose appalling living conditions only became worse when the winter snows arrived. They also worked on drainage as it was feared heavy winter rains could wash away defence works and break down trench walls.

A new aspect to the Gallipoli fighting left many of the miners' existing works in tatters: the arrival in late November of four heavy Austrian siege howitzers. Beginning on the morning of 29 November, the Turks shelled the Australian positions at Lone Pine, the shells exploding four metres below the surface, shattering tunnels and galleries and collapsing trenches. Suddenly massive holes appeared eight metres in diameter, large chunks of debris were thrown into the air and men were blown apart and buried.

The evacuation began soon after, on 19–20 December. Carre Riddell, the officer in charge of the mines opposite Johnston's Jolly, wrote late on the last day: 'It seems very sad to give up the work now after all the lives and money it has cost, but we had to realise that we could not have got through without sacrificing ten times more than all which had gone before and that we were needed elsewhere.'[1]

On the Western Front, the formidable John Griffiths was working on the most ambitious mining plan ever conceived. It would tear apart the Messines Ridge and alter the course of the war, and Australian tunnellers were to play a crucial role in its success.

✕

The Germans had taken the lead in offensive mining on the Western Front, but week by week the number and strength of the British tunnelling companies had increased, driven on by the passion and enthusiasm of Griffiths. The first five tunnelling units, numbers 170 to 174, were rushed into the fighting. But companies 175 and 176 instead spent time first at a special training area near their billets behind the line.[2] Unlike the early haphazard use of mining companies, there was a new focus on offensive mining. Once they had finished their training, the 176th Company was ordered to begin mining near the Orchard at Bois du Biez,[3] while the 175th Company received orders to start tunnelling at Railway Wood, Hooge and Armagh Wood. One of their sites was just 2.5 kilometres away from Hill 60, on the northern side of the Menin Road, at Hooge, a large property with a chateau and a number of outbuildings, including stables and a coach house. It was a dangerous area as the frontlines were very close, the Germans holding a strong defensive line within the ruins of the chateau, while the British line ran through the stables. High Command had given orders to drive the Germans out. The British also wanted to use the attack as a diversion – or, in the terms of the time, a 'demonstration' – to draw the Germans' attention away from an upcoming attack at Loos.

In mid-June 1915, volunteers were sought to undertake the perilous mission of mining the site. After a few shots of apricot brandy from the local brigadier general, Lieutenant Geoffrey Cassels volunteered. Cassels and his CO, Major Hunter Cowan RE, set out from Ypres a few days later to conduct a recce of the British frontline. The land here was slightly higher than the surrounding countryside, so it had been continuously fought over since late 1914. The ground was strewn with the dead; the corpses were black and rat-eaten and gave off a nauseating smell.

The two officers looked for a suitable site to start their tunnelling. They needed somewhere close to the frontline, to minimise digging, but far enough away to be reasonably safe. Looking around, they decided on the shattered stables as their thick walls would give some level of protection and also hide the tunnel entrance and mine workings.

Cassels was placed in charge of the operation. He was provided with 40 men from the 175th Company. Many of the men were new to the front, having just been recruited, kitted out at Chatham and sent over. So Cassels had a job on his hands for the first week or so, getting the men used to the horrendous living conditions and the battle that raged around them. Unlike most units in the Allied army, the miners were issued copious quantities of rum sent up in stone demijohns bearing three simple letters: SRD, meaning 'Strong Rum – Dilute'. They certainly needed it.

The tunnellers soon encountered problems. Instead of solid ground, they found wet running sand, virtually impossible to dig through and contain. They drove piles but found that their pumps could not cope with the sand and water. To try to contain the wet sand, they stripped the kilts from dead soldiers of a Scottish Highland regiment and stuffed these behind the piles, to no avail.

Cassels set out to find a more suitable location to begin work on the new shaft. German snipers were active and, with the two frontlines so close, movement was extremely dangerous. He made it to the gardener's cottage on the estate, where he found a deep cellar and an ideal starting point for the shaft. The ground here was drier and the cellar afforded protection. He quickly moved in his men and had them begin digging. Forcing the pace, the men sank a ten-metre shaft and hit blue clay, much to the delight of the clay kickers.

The Germans were in the midst of constructing two concrete redoubts, and the ruined chateau now held few German troops and had become more of a sniper's post. It was decided that the tunnel should be redirected towards the concrete redoubts rather than the chateau as originally planned. Cassels was given three weeks to drive a 60-metre-long tunnel beneath the larger of the two redoubts, plus a branch tunnel of 30 metres under the other.

The tunnel started out at roughly two metres wide by one metre high, but to increase the pace it was soon reduced to 1.2 metres by 0.6 metres. While the clay kickers sweated at the face, men quietly dragged out the spoil, rolling it back on small trolleys that ran on wooden rails. The clay was then bagged, dragged to the surface and taken back behind the line. It was camouflaged so that the region's distinctive blue clay would not give the enemy a hint of the depth or location of the mining operations.

Within the shaft, the air was dank and difficult to breathe. They had tried using mechanical bellows to pump in fresh air, but the miners worried that the rhythmical *woofing* would alert the Germans. It also made listening for German countermining difficult. Much like Hill 60, the ground was perforated by shafts and galleries, and German tunnellers were often working just metres away. Lieutenant Cassels could be found alone at the face, listening and gauging the location and direction of the German tunnelling. He lay with his face pressed into the clay, straining to hear any movement in the earth that would betray the enemy. The atmosphere was stifling, claustrophobic and frightening – a few inches above his head, the damp clay, and above that, ten metres of heavy earth weighing down upon him.

Apart from the tiny area lit by the flickering candle, around him was silence and darkness. Cramped as he was, he could feel the wet, cold earth all around him, just above his head and tight on his shoulders. And the ground seemed to murmur and move and close in on him. It was so very, very close. Movement was difficult. He lay prone on the clay floor of the tunnel, his head pressed against the face, listening, listening. His heart thumped and often it was hard to differentiate this from those other diverse and confused sounds around him. As he says in his diary: 'Some sound could be heard, dull and muffled. There was always that fraction of a second of doubt – where it might be enemy mining. One's pulse rate quickened and fright pushed to the fore one's whole being.'[4]

While the main tunnel drove forward and under the larger concrete redoubt, the branch tunnel, due to the poor calculations and primitive survey equipment, was found to be in the wrong position for laying a charge that would destroy the smaller redoubt. With no time to dig a new branch tunnel, Cassels had a serious problem, one that could only be solved with a 'bold and risky' action, as he put it.[5] If he could not destroy the second redoubt directly, perhaps he could bury it in debris – but for this he needed a massive explosion.

He found the answer in a little-known explosive charge called ammonal, never before used by the British army. All Cassels really knew about it was that it was composed of 75 per cent ammonium nitrate, with aluminium and TNT, and had a rate of detonation of five kilometres per second and over three times the lifting power of gunpowder. However, he had no real idea of the capabilities of the explosive given the depth of his tunnel and the weight and composition of the German redoubt. He calculated that he would need about 1.5 tonnes of this mystery substance.

The first problem was to get a supply of it. Cassels put in the necessary paperwork to the quartermaster general at GHQ, and his request quickly became an embarrassing issue. 'Ammonal' was confused with the drug 'Ammonol', used extensively in America 'as a sedative in cases of abnormal sexual excitement'.[6] You can imagine the poor quartermaster thinking that if they needed over a tonne of the stuff something strange was certainly going on along the front among those troublesome unionist miners.

The confusion was eventually straightened out, and the quartermaster dispatched the ammonal to the front. But when the ignition deadline was only three days away, there was still no sign of the ammonal, and Cassels raced around seeking alternative explosives for his tunnel. He received a signal to say his ammonal had been dispatched and should have arrived on a wagon already. Concerned at its delay, he frantically searched for the shipment. It was now daylight so a slow-moving wagon, whatever its cargo, was fair game for the ever-vigilant German gunners. At the last moment, the wagon appeared out of the low ground fog, to the great relief of Lieutenant Cassels. The driver explained he had been delayed because he'd had a problem with one of the wheels.

The ammonal was unpacked and lugged into the mine, down the narrow, slippery tunnel, and stacked in the dark, grim blue-clay gallery. The explosive was in 70 tin containers sealed with pitch, as it was very susceptible to moisture. The tins were stacked up like a mountain and spread among them were 24 detonators in batches of six. Each detonator was about 2.5 centimetres long and half as thick as a finger. These were connected to two sets of electrical firing leads, or fuses, that ran back to a communication trench. When the handle of the exploder was depressed in the communication trench, a current would pass down the firing leads, the detonators would explode and the charge would ignite the tins of ammonal.

Cassels worked quietly and diligently in the trench, testing and re-testing the circuit with a weak torch battery. The assaulting troops shifted nervously on their start tapes in nearby trenches and shell holes, apprehensive about the size and fallout from the explosion. The tunnellers retreated to a safe distance, prepared themselves and watched the seconds tick down. All was ready.

Suddenly, a German heavy-calibre shell crashed into no-man's-land, throwing dirt, barbed wire and timber high into the air. Horrified, Cassels tested the line – there was no reading. It was minutes before zero hour, and after all the work, the pain, the fear, the great mine was dead. He tested again and the circuit was still dead. Frantically Cassels, along with another officer and a corporal, ran back into the explosive-packed tunnel. He could imagine the assaulting troops sweating anxiously as they huddled low beneath their parapets, trusting that his mine would obliterate the Germans and thus save their lives, but right now there was no mine, no explosion, and zero hour would be their end.

His hand found the firing wires in the yellowish glow of the torch beam. Next to him, the corporal ran these wires through his fingers, following them towards the German line. He found the break, clean and neat across the two electrical firing wires, and quickly set to work crudely joining the ends together. They raced back along the damp, timber-lined tunnel to the shaft, climbed the steep wooden ladder and gulped in the clear air above. Again they tested the circuit. The faint continuity test showed that the repair had worked. They were back in business.

Cassels got into firing position, his exploder in his hand. He looked at his watch. There were just over four minutes to zero hour. He watched the hand slowly tick by and wondered at the birds singing in the peaceful dusk of summer. Not 100 metres away, the Germans were beginning their dinner. He'd caught the smell of fried meat, of sizzling fat and the aromatic tobacco his enemy so often enjoyed. Just a pity about the ammonal lode beneath their feet.

At exactly 7 pm his plunger went down. The massive mine exploded. High into the air went the trench line of the Germans: trees, timber, barbed wire and tumbling bodies. No sooner had the debris landed than German artillery, well aware of the attacking troops, opened up. Shells blasted the British line as Cassels ran back along a narrow communications trench, past the wounded attacking troops, and made his way to the relative safety of Sanctuary Wood.

The explosion had been massive, so massive he wondered whether the frontline of the attacking British troops had been buried along with the Germans. He soon found out that ten men of the 4th Middlesex who were positioned in an exposed trench had been buried by the falling dirt and debris, as had a forward ammunition supply dump. This would require explaining.

Sure enough, he had hardly laid his tired head down on his smelly, lice-ridden pillow when he was awoken and dragged, stumbling, into the sidecar of a military-police motorcycle. As the cool air streamed across his face, he wondered at his fate, knowing the loss of soldiers caught in his blast could easily be sheeted home to a lowly mining lieutenant. Dazed and exhausted, his uniform caked with mud and his face unshaven, he was paraded before Major General H. de B. de Lisle, who angrily questioned him about the blast and informed him he was under arrest. Cassels had little to say, other than that he thought he was doing his job and had no control of the men who had been positioned so close to the blast.

Suddenly the door behind him opened, and in strode Lieutenant General Sir Edmund Allenby, the commander of the 5th Corps. Cassels was dismissed and sent outside. Soon after, he found himself sitting in Allenby's staff car, being driven off to lunch somewhere with Allenby and Commander-in-Chief Sir John French, who warmly congratulated him on a great effort.

By the end of the day, he had been awarded a Military Cross and granted ten days' leave in England. Amidst the excitement, his threatened court martial was somehow forgotten.

✕

In all, it had been a great day for the tunnellers. More importantly, Cassels' bold mining operation was to become the inspiration for John Griffiths' grand plan for a series of explosions that would rip Messines Ridge apart and provide the breakthrough that the Allies so desperately needed. At the time of the Hooge blast, the Western Front, particularly the staunchly defended Messines Ridge, was still in a gruelling stalemate. Allied attacks at St Eloi had failed, and the ridge protruded dangerously into the Allied front.

Griffiths, in his usual far-sighted and visionary way, thought that what was needed was a well-planned and concentrated attack beginning with a series of massive mine explosions all along the German front. He had been impressed with the effect that Lieutenant Cassels' mine at Hooge had on the German defenders and believed that if the operation was planned right, the ridge, and certainly the protruding salient, could be virtually flattened out.[7] He put the idea to his immediate superior, General Harvey, suggesting that six deep mines be fired by his tunnellers.

General Harvey was impressed and agreed to present the suggestion to the engineer-in-chief, General Fowke. However, the overworked Fowke immediately rejected the plan and cursed Harvey and Griffiths for subjecting him to such a preposterous idea. How could the British tunnellers mount such an attack when the Germans were running rings around them? And anyway, defensive tunnelling was more important. So the idea was shelved.

Griffiths was not one to be put off. With Harvey's support he eventually brought General Fowke around to the idea late in 1915, but they still needed to convince the British High Command, who were cool on the idea but respected Fowke's experience. By the end of the year, the idea of a massive mining attack started to gain traction. In early January 1916, Fowke, Harvey and Griffiths were ordered to attend a high-level conference at GHQ and discuss their idea further.

Faced with an audience of grey-headed old men and generals sporting red tabs and braid, Griffiths was in his element and addressed them with his usual enthusiasm and vigour. Here, he said, was the ideal weapon to smash open the German front, break through, and in the process save 10,000 Allied lives. His tunnellers would create what he termed 'an earthquake' that would destroy the ridge, swallow the German garrison and shatter the enemy's nerves.

The generals looked on incredulous. Who was this pompous man, a mere major, with such outrageous ideas? And who was he to speak so forcefully to them? Did he have no respect? Given the litany of frontal attacks that had failed – even those with the support of underground mines – such a 'breakthrough' did not seem feasible, and the logistics looked impossible. What was next on the agenda?

Well, the trio had tried, they had put their case. Griffiths had been given the chance to show his best dance steps, but to no avail. High Command rejected the scheme, and so they returned to their billets in a deep state of rejection and despair. Old ways and old thinking were still prevalent, and the concept of a new, aggressive mining operation to break the stalemate was simply before its time.

Late that night, a dispatch rider arrived at General Fowke's quarters with a message marked 'Urgent'. Tearing open the manila envelope, Fowke's eye quickly scanned the lines – it was an amazing turnaround. The High Command had thought about the idea and now saw merit in it. They had changed their minds. Fowke raced across and woke Harvey and Griffiths and blurted out the news: they liked the idea of an earthquake, he said, and the tunnelling attack on the Messines Ridge had been given the green light. The tunnelling companies were back in business in a big way.

Fowke, Harvey and Griffiths' presentation to High Command had been serendipitously timed. The Allies had already decided they needed to mount a major offensive: something really big that might blow open the front and win the war. Planning was already secretly under way for what would become the Somme offensive of July 1916, and if there could be a parallel offensive in the north, along the agonising Ypres salient, then there was value in the Messines idea. If it did allow a breakthrough, perhaps the British could carry the offensive to the coast, an enticing possibility.

SEVEN

Along the
River Somme

While Griffiths was developing his plans for a bold mining operation along the Ypres salient, more new British tunnelling companies were being formed, and they joined existing ones to begin the takeover of the French line north of the Somme. Here the mining conditions were very different from those of Belgian Flanders. The trenches abandoned by the French were full of their unburied dead, often two and three deep. They were shallow and in some places only 30 metres from the Germans, making them extremely dangerous. Instead of running sand and blue clay for the tunnellers to dig down into, there was chalk. On the plus side, it required far less internal support. But while clay allowed the tunnellers to dig silently, tunnelling in chalk was noisy, and neither side could easily mask their operations. It also increased the risk of gas poisoning as the carbon monoxide produced after mine explosions seeped into the porous chalk and lingered. On average, three men would become gas casualties each day.

Carbon monoxide poisoning is a silent killer. Being colourless, odourless, tasteless and not causing any irritation to airways, eyes or skin, it is difficult to detect. A low dose for six to eight hours leads to vertigo, disorientation, headaches. More serious exposure brings severe headaches in one to two hours, and a high dose means headaches, nausea and dizziness within five to ten minutes, and death within half an hour. Men often staggered from a gas-filled mine shaft and collapsed on the surface. Others, overcome by the gas, could not climb ladders and died at the shaft bottom or fell from the ladders, too weak and disorientated to climb to safety. Even those who did make it to fresh air were often hospitalised for long periods suffering from memory loss, depression, severe head pain and confusion as the toxicity in their system affected their heart and their nervous systems.

Men were also affected in small, airless shelters and dugouts, as wood heaters, petrol-powered equipment and portable stoves all created carbon monoxide gas. This was especially the case in winter when draughts into shelters were blocked or covered by snow. The men had only the crudest ventilation and exhaust equipment in the tunnels; some said it was left over from the Crimean War in the 1850s, so they used caged mice and birds as their early warning system to detect gas.

To protect themselves from carbon monoxide poisoning, rescue teams and tunnellers working in dangerously poor air quality wore Proto kits. A Proto kit was an apparatus that consisted of two cylinders of oxygen and a large re-breather bag with separate inhaling and exhaling compartments, from each of which ran a tube up to the mouth. The cylinders of oxygen were carried on the back and the re-breathing bag lay on the chest, giving rescuers freedom of movement, and as there were no parts of the apparatus that projected out to the side, the kits could not be damaged in the confines of a tunnel. The rest of the kit included a pair of smoke goggles worn over a skull cap, and a nose clip that forced them to breathe only through the mouth, enabling them to work for up to two hours in safety.

The Proto kits were developed from an apparatus used on the English coalfields. In June 1915, six Proto sets were dispatched to the Royal Engineers to assess their suitability. Two experienced mine-rescue workers, Corporal Ellison and Lance Corporal Clifford, trained 12 tunnellers to operate the equipment. The kits were approved for use, and they extended the training to the other tunnelling companies. By the end of 1915, they had trained nearly 3000 men. Their work resulted in the formation of the Mine Rescue School at Armentières.

The British tunnellers on the Somme trialled some new tactics. At the Old Mill mine near Cuinchy, they fired two small mines to draw the Germans to the frontline in anticipation of an attack, but these small explosions were no more than bait. As expected, the Germans crowded into their firing line, just in time for the much larger charge to be fired. This smashed a long section of trench, killing many of the enemy garrison and wiping out the defences in this section of the line.

In late July, the 174th Company established a headquarters at Bray-sur-Somme and took over 66 French shafts between La Bois-selle and Maricourt, including some at Carnoy and Fricourt. The dashing, handsome grandson of the Duke of Wellington, Captain Edward Wellesley, established his headquarters just north at Méaulte near Albert, and his newly formed 178th Company began works in the French tunnels at Tambour Duclos, near Fricourt. The 179th, also recently formed, began work at La Bois-selle, just to the northeast of Albert. This concentration of mining companies began the work that would see the explosion of nine mines on the opening day of the Battle of the Somme, including the massive Lochnagar mine, the crater where many visitors pay their respects today.

Wellesley's orders were clear: the 178th Company was to destroy the enemy tunnels and workings, undertake offensive mining against the German positions at Fricourt and hold the high ground at Tambour. From the time the British tunnellers took over the maze of French workings, the sound of German mining was clear as they aggressively pushed towards the British line. A deadly game of cat and mouse developed, as each side dug forward, listened, fired a camouflet and drove on. It was when the sound of the enemy's digging stopped that the period of real danger began, as this was when the explosive charges were laid, the tamping completed and the mine fired. Officers needed to interpret the sounds and withdraw their men quickly, before the enemy fired their mine.

In early 1916 the Germans blew a camouflet at Fricourt that collapsed a long section of the Allies' tunnel. In such cases, men usually died quickly, either from the percussive force of the explosive or from the carbon monoxide gas, but on this occasion tapping was heard. A rescue team led by a young Canadian second lieutenant, Robert Mackilligin, raced to the collapsed section and immediately started to dig the men out. But the chalk was saturated with surface water, and keeping it from collapsing meant the rescuers' progress was slow. Finally a small hole was bored through and, by the light of a torch, two men could be seen, face down in the narrow collapsed tunnel, their feet trapped.

Suddenly, from behind the trapped men, thick, chalky water began to trickle over the collapsed tunnel roof, which had formed a low dam. Frantically the rescuers dug forward as the water swirled around the trapped men. Mackilligin felt the chalk begin to close in on him, too. He had to get out. In the faint beam of his torchlight, he saw the two desperate trapped men as they tried to lift their heads clear of the thick, chalky water, until it submerged them. They gulped in the white slurry and died a frightening death.[1]

Also working near Fricourt, at the notorious Tambour Duclos, was Lieutenant Edmund Pryor, a 21-year-old student mining engineer who, with a year's service behind him, was considered an experienced soldier when he reported for service with the 178th Company. He was an unusually tall man at six feet two inches – the average at this time was about five feet six inches – and broad-shouldered, which gave him problems in shallow trenches and narrow tunnels. Late in 1915, Lieutenant Pryor squashed his tall frame into the end of a forward tunnel and placed his ear to the chalk. He could clearly hear the Germans tamping their mine, pushing the bags of chalk into their tunnel to maximise the force of the explosion towards the British workings. There was little time before the explosion would rip through his tunnels, so he had to move fast. Quickly he ordered his men to drive an untimbered tunnel forward at a frantic pace. It was a little over half a metre square, about the same width as a coffin. When they had gone about 14 metres, Pryor squeezed himself into the narrow tunnel feet first and, lying on his back, wriggled towards the face. He could feel the chalk inches above his nose and wondered if this would be his tomb.

Pryor's men passed heavy 50-pound bags of gelignite to him, which he pulled across his body and pushed into place at the end of the gallery with his feet. He pushed bag after bag into place until a load of five tonnes was prepared for firing. The firing wire trailed back to the shaft. Pushing the final tamping into position, he scrambled from the mine and retreated along the tunnel to the surface where, covered in chalk dust and dirt, his shirt stained with sweat, he climbed on his motor bike and raced back to the headquarters in Méaulte to ask Captain Wellesley for permission to fire his mine. Wellesley refused, insisting that the firing be coordinated with a possible surface attack, which had not been done.

The following morning the Germans fired their mine, killing three of Pryor's men. Their bodies were virtually atomised, and all he later found was a shattered hand sticking out of the tunnel wall.[2] The loss of these men was just the tip of the iceberg. By the end of 1915, the 178th Company had lost more than 300 men.

On another occasion, Lieutenant Pryor was climbing down a shaft in an attempt to rescue trapped tunnellers after the Germans exploded a camouflet and he fell nearly 25 metres down the shaft, smashing into broken timber before landing at the bottom caught up in his rescue equipment. He blacked out, then came to long enough to realise he had been hauled out and was lying on the floor of a trench. Slipping back into unconsciousness, he lay there battered and bloody until he felt hands roughly lifting him and throwing him high over the back of the trench and onto a pile of bodies.

He lay there freezing in the bitterly cold December sleet with nothing on but his trousers and one of his gumboots. Suddenly, high above, a shrapnel shell burst and the ground around him came alive with the deadly patter of shrapnel balls, one striking him. He could feel death coming, and again unconsciousness overtook him. But Pryor was lucky. His loyal batman came to search his body for a memento to return to Pryor's parents and was astonished to see the slight flicker of an eyelid – he was alive.

Pryor may have survived, but for him the war was over. He was invalided back to England, where it would take him many years to regain his strength and health.[3]

Tales of extraordinary heroism and tragedy were being played out across all the tunnelling companies on the Somme. In late 1915, Lieutenant Eaton of the 184th Tunnelling Company, stationed near Suzanne on the front to the south of the River Somme, had wondered what was going on as there seemed very little activity in the area. As Lieutenant Eaton saw it, the only way to know about the German mine workings was to go over and find out.

On 4 January 1916, he headed out, the overcast sky and the new moon giving little light, which suited him well. He was already cold as he slipped quietly out of his dugout. Away to the north, an Allied strafe was hitting German reserve lines up near Fricourt, where Wellesley's 178th Company was sweating and hard at work underground.[4]

He slipped through the wire and splashed out into the marshes that protected this part of the front. Soon he found himself on the edge of the River Somme. Winter rains had lifted the river level and had strengthened the current, while war debris and waste from upstream had tainted the water. Fully clothed and as quietly as he could, he slid down the bank and into the icy water. He had greased himself thoroughly with pig fat before he set out, but the chill gripped his throat and took his breath away.

For ten minutes he swam noisily across the river, his body vertical because of the weight of his boots and his clinging heavy woollen uniform. At his waist he carried a pistol and a small pouch of loose rounds, in his pocket he had some Mills grenades, and along his web belt he had a solid block of timber he'd fashioned into a trench weapon. By the time he'd swum across the current and dragged himself up the muddy bank, he had travelled unchallenged well over a kilometre behind the German lines, but he had no real idea where he was.

Dripping and cold, he headed inland, splashing through the shallow marshland that bordered the river. Since the Germans had routed the French from the area they had neglected it, patrolling it only infrequently. Travelling north, Eaton found a German track and followed it through the marshes and around the flank of the German line.

He came upon a German ammunition dump guarded by a single sentry, whom he quietly overpowered. But the noise had disturbed other German sentries, and suddenly men seemed to emerge from everywhere out of the darkness. Nearby was a wagon, its load covered with a heavy tarpaulin. He slipped in and remained still, his heart pounding as the Germans began their search. To his surprise, two men climbed onto the wagon, and it began to drive off. Quickly realising it was going in the wrong direction, Eaton crawled towards the rear and very quietly dropped over the back and onto the soft, muddy road.

Getting his bearings, he turned and headed south. Close by in a wood, a German gun crew were firing on a fixed bearing, perhaps a British road junction or a known assembly point. He noted the location of the crew and moved on. Soon he came upon a German reserve trench and moved quietly along behind it in the darkness. Still he was not noticed. Coming to the rear of another trench line, he saw a faint glow from a dugout below. Pulling back the hessian curtain, he lobbed in one of his Mills grenades, which exploded a few seconds later. Still nobody saw him or even came to investigate.

Eaton quickly moved down into the trench and away from the commotion, keeping an eye out for any sign of mining, spoil dumps or the entrance to a shaft. There were none, just as he had suspected. Ahead, he heard some muffled orders, men shuffling on the wooden duckboards as they collected around an officer before forming into single file and heading away down the trench. It was obviously a patrol about to go out into no-man's-land. Eaton tacked himself to the end and headed out into the darkness with them.

The patrol moved quietly towards the British line with Eaton in tow. When he thought he had chanced his luck enough, he dropped silently to the ground and waited until the Germans had moved on. Looking up, the sky was slowly brightening with a cold dawn light, so he continued south until he again found himself at the Somme. He pushed out into the freezing water and swam to the far bank, to a section held by the British. He climbed the bank and made his way back to his muddy dugout where he gulped down a mouthful of issue rum, climbed into his bunk and slept soundly.

Later that morning, after writing his report and noting the position of the German guns, he was summoned to GHQ. There he was paraded before the commander-in-chief. After initially being asked why he had disobeyed orders, risked his life and potentially given intelligence to the enemy, he was warmly congratulated on his initiative and bravery, and was awarded on the spot the Distinguished Service Order, one down from the Victoria Cross. Soon a French Legion of Honour and the Croix de Guerre with Palm followed and, more importantly for Eaton, a move to a more active section of the line where his skills and his keenness could be put to good use.[5]

✕

Through the latter part of 1915 work also continued apace under the frontline around Hill 60 and down along the Messines Ridge. There, too, bravery was the order of the day.

It was inevitable that the British and German tunnels would

sometimes meet and intersect. Perhaps it would begin with a cave-in that revealed a black hole in the face or wall of a tunnel, or the tunnellers would suddenly feel a faint breeze coming from a hole through to the enemy's tunnel, or the rear of the enemy's lining timbers would be revealed suddenly.

On 9 November 1915, British tunnellers of the 172nd Tunnelling Company near the Bluff, just 1500 metres to the southwest of Hill 60, were carefully boring towards the enemy lines when they struck German mine timber and called Second Lieutenant Richard Brisco to investigate. After enlarging the hole with a bayonet, Brisco wriggled forward, removing some lengths of lining timber, and dropped into the German tunnel. Taking his pistol out, he moved cautiously forward in the dark. A group of German tunnellers emerged from the darkness just metres away. Immediately Brisco fired and a German dropped to the ground. The rest turned and ran. Brisco turned, too, and slipped back into his own tunnel, where he grabbed a small portable charge, lit the fuse and dropped it into the German gallery, sealing off the tunnel and bringing down the roof.

In late December 1915, Brisco, who was making a name for himself in underground exploits, entered a German tunnel, this time in the side of a mine crater after a small land slip. Again he met up with a party of Germans, raced back to the entrance and began firing. A short, sharp fire-fight ensued, until Brisco barricaded up the entrance with a few sandbags of earth and raced back to his lines.

In February 1916, a listening post reported a German tunnel being dug towards the British line. Brisco immediately started his own tunnel to intercept it, then charged and tamped it. When it was reported that the Germans had broken into his gallery and were removing the tamping, he raced into the mine. He cleared away the bags of earth and, when there was only one layer left between him and the Germans, waited with pistol drawn for them to break through. A hole appeared, and then a face. Brisco fired and the German dropped. He pushed the dead German aside, climbed into the tunnel and fired a camouflet, bringing down the roof and thwarting another enemy advance.

The following day, Brisco was again in action, crawling out into no-man's-land to drop into the entrance to an enemy tunnel that had been spotted in a crater. Twelve metres in, he rounded a corner and a German machine gun opened up on him. Diving for cover, he pulled a grenade from his tunic, whipped out the pin and sent it spinning along the tunnel towards the Germans. The explosion in the close confines of the tunnel was ear-splitting. As the smoke cleared, Brisco leapt up, jumped over the dead gun crew and raced along the tunnel towards the German lines. At the bottom of the German shaft, he placed a small portable charge, lit the fuse and raced back to the tunnel entrance. He emerged back into the night, the darkness now lit by flares, but was wounded along with other tunnellers when a German machine gun caught them in the open.[6]

In the days after, the Bluff was blasted with artillery and heavily mined by the Germans as a diversion from major offensives against the Allies at Verdun and Vimy. The Germans captured the Bluff, with 67 British officers and 1227 other ranks killed or wounded. The tunnelling companies also suffered, with nearly 50 men listed as casualties. The British mounted counterattacks and drove the Germans from the high ground, but it was with tragic losses that highlighted just how costly the fight for these strategic positions continued to be.

EIGHT

Misspent Energy
and
Wasted Effort

As 1916 opened bleak and freezing, there seemed little hope of victory in the foreseeable future. There had been no progress on the front, and the army that had taken the field in 1914 was virtually gone, with nearly 400,000 casualties in slaughters that had achieved very little. The current force, Kitchener's Army, was a conscript army that was only half-trained and led by ageing generals who, even after the battlefield disasters, had little appreciation of the new machines of war or what tactics to use against them. In England, people felt humiliation, even mortification, about the course of the war so far, and many were deeply pessimistic about the future.

The British tunnellers' mining activities for 1915 were considered unsatisfactory, and it was accepted that in mining the Germans certainly held the upper hand. Despite their bravery, sacrifices and hard work, Britain's efforts had produced few tangible outcomes. The contribution of the tunnelling companies was being closely scrutinised by High Command and their value reconsidered, for they were absorbing large numbers of men who might have been better deployed fighting the Germans on the surface rather than underground. The tunnelling companies were demanding equipment, stores, metal fabrications and explosives, all of which could arguably be put to better use. And the mines that they did successfully blow more often than not created vast craters that obstructed the attacking troops, forcing them to bunch and providing attractive targets for German machine guns. The Germans would quickly put the craters to good use and successfully defend them, often leaving the British in a worse position than when they set off the mine.

In the British tunnelling companies, many of the officers had little knowledge of what was needed, and were inexperienced in mining – certainly in military mining. Captain Grieve in his definitive book, *Tunnellers*, wrote: 'A major part of the work must be written off as misspent energy and wasted effort.'[1] Perhaps had the most senior tunnelling officers – General Fowke, Colonel Harvey and Colonel Edmonds – been in the field to actually direct operations rather than 50 kilometres behind the front at GHQ, things might have been different. Griffiths had also been busy, racing around the front in his muddy Rolls-Royce, recruiting in London, stirring up the men and being the link between the tunnellers in the line and the far-off staff at GHQ.

And so, from the British High Command down to the men in the tunnelling companies, 1915 was put down to experience, to experimenting and learning lessons in an environment they were completely new to and certainly untrained for. They looked to lift their game and move decisively from defensive to offensive mining. To improve communications between the tunnellers and GHQ, Colonel Harvey RE was appointed to the new staff post of Inspector of Mines and under him were three Controllers of Mines for each of the 1st, 2nd and 3rd Armies. The British re-organised their tunnelling companies under the overall control of

GHQ in late 1915 and into early 1916, so that all the British tunnellers spread along the front were controlled by one authority. Better overall planning saw the failures of 1915 corrected.

This reorganisation was aided by improvements and standardisation of their equipment, the introduction of more reliable and quieter pumps and ventilation equipment, rescue equipment, more accurate surveying tools, safety procedures, geological processes, enhanced training and sensitive listening devices.

The latter were crucial because the primary method of preventing the enemy from breaking into the Allies' galleries and workings was by listening for sounds of their tunnelling. It was difficult just to pick up the sounds of the enemy, who were trying to tunnel as quietly as possible. Even more complex was calculating the approach, angle and speed at which an enemy gallery was being dug.

In the early days of the war there had been no listening devices at all, and men would simply press their ears to the ground or against the cold, damp face of the tunnel and strain to hear any noise coming from the earth. Then men began holding a length of stick against the face or the wall of a tunnel, using this to focus sound and detect slight tremors. Or they drove a spike into the ground and attached it to an electrically operated tuning fork and diaphragm. Griffiths had approached the Metropolitan Water Board, who supplied him with listening sticks, which they used to detect water leaks from their mains. They were simply short sticks with an attached vibrating wire and earphone.[2]

A popular technique was developed by a French soldier using two army water bottles filled with water and laid on the ground side by side. Sound was transmitted through the water, and so when the men pressed their ear or a medical stethoscope up to them, they could hear better. Of the many devices tried, this was considered the best, but none were reliable as it was difficult to distinguish the sound of enemy tunnelling from shelling, trench activity, trickling water and from the natural creaking and movement of the ground.

And then, along came the geophone, which Griffiths discovered in December 1915 by accident while looking for other equipment for his tunnellers. The geophone had been developed at the Sorbonne, based on the water-bottle technique. It consisted of a pair of wooden discs about ten centimetres in diameter by four centimetres thick. Inside each wooden disk there was mercury contained between two mica plates. The discs were connected by rubber tubes to stethoscopic hearing pieces, and the listener placed the discs on the ground and knelt between them. Sounds were transmitted through the discs into the earpieces, and the listener moved the discs around until the sound coming through each ear was the same volume. The direction the sound came from was perpendicular to a line between the two discs. The listener took a compass bearing on that point and wrote down the reading. Bearings could be taken from the faces of two or more other tunnels, and all the bearings plotted on the same piece of paper. The enemy's location was where the lines intersected.[3]

Another device was the seismomicrophone, basically an electrical sound detector that could be placed at the end of a tunnel and connected by wire to a switchboard. A number of these could be connected to cover a whole section of the line, saving manpower and the exposure of listeners to the risk of countermining. However, they were nowhere near as effective as the geophone, and in some instances hardly better than an ear pressed up to the face.

The accuracy of geophone readings depended upon the listener's skill and experience. To be a listener it also took a special kind of person who could sit cramped at the end of a long, lonely tunnel for hours – a tiny candle the only source of light – with the ever-present fear that the Germans would explode a mine. All the while the listener had to keep a cool and focused state of mind. He needed good leadership, decent periods of rest out of the line and vast quantities of rum.

The listener's procedure was well rehearsed and efficient. On hearing what he perceived as German underground work, he immediately reported it to his duty officer, known as the trench officer or simply the 'trench'. The officer would return underground with the listener and crawl into the tunnellers' listening post and take up the geophones. If he confirmed the sound, other listening posts would be informed and they too would listen for the same sound. Bearings would be taken and plotted on the mine map, and this would continue over the next few days or weeks, to determine the progress of the enemy tunnel. They could then calculate accurately when the Germans would be closest to their own tunnel without breaking through, and when and where it was best to explode a camouflet.

To give the listeners every chance of detecting the Germans, it was standard practice to cease tunnelling work at regular periods to give the listeners a period of silence. They also ceased work, except in one section, to check the volume of their own tunnelling, so that the listeners could establish if they – and the Germans – could detect any noise.

The infantry manning the frontline trenches were reassured to know that up-to-date listening equipment was being used below them. They were quick to call in the tunnellers if they had the slightest suspicion that German mining work was going on below them, so tunnelling officers were continually being called out to investigate noise. Sometimes these were false alarms that took mining officers away from their real job. Baby mice squeaking, rats scurrying, water dripping, men hammering, splitting kindling or digging extensions to funk holes, and even the croaking of a frog had at various times been mistaken for German mining activity.

Lieutenant Cassels, who had come to prominence with the Hooge ammonal firing, found himself continually annoyed and distracted by the need to investigate reported noises. When he was certain that it was a case of jitters and imagination, not German mining, he would turn up with an impressive 'listening device' he had knocked up himself, a box bristling with wires and connected to a stethoscope, which he proceeded to move around with a certain level of dramatic effect. This and a reassuring chat were usually enough to comfort the men and soothe their nerves.

Another British tunnelling company, when called upon to investigate what was reported as German tunnelling, designated two ageing, near-deaf, ex-coalminers to check. They would turn up, much to the relief of the frontline troops, and solemnly go about their task. After listening and going through some well-rehearsed theatrics, they would say, 'Yes, we can hear them, all right. They're there. It sounds like they're f***ing.'

Astounded, someone would usually ask, 'What, do you mean the Germans?'

To which they would wryly say, 'No, the rats.'

This became a well-known joke all along the front and even the Germans were said to have heard the story.

✕

As a result of Griffiths' work and his recruitment office in London, there were now 20 tunnelling companies in the field, made up not only of specialist miners but also engineers, mine managers and even doctors from the mining districts of Great Britain who were familiar with miners' diseases and rescue work.

The miners may have been experienced and enthusiastic, but they were never your typical, smartly dressed, British parade-ground soldiers. The tunnelling companies were taking on their own persona and their own independent spirit. Many of the men had spent a year at the front and as their work was very different and their skills unique, they had evolved from being Royal Engineers to virtually a private army engaged in a different war in a very different space to the men above ground.

Tunnellers were on occasion hated and their presence shunned. When a mining company turned up and started work, the troops nearby often became edgy and nervous, fearful of drawing German tunnellers to their quiet patch of the front. But when the enemy could be heard below the frontline, the soldiers had a new sense of respect for the tunnellers. They were made very welcome then, and the troops were quick to offer their moral support, their muscle and their protection. General Harvey stated: 'The men would not stay above unless the miners were beneath. That is a fact.'[4]

The first honours for effective destructive tunnelling in 1916 went to the Germans with the ignition of the New Year's Day mines south of the La Bassée Canal, to the east of Béthune. The British retaliated on 2 March with the exploding of three mines under the German frontline at the Chord, near the Hohenzollern Redoubt north of Loos, an area of incessant and bitter fighting.

But it was still the high points along the line near Messines – St Eloi, the Mound, the Bluff, the Caterpillar and Hill 60 – that drew the attention of the planners. Now the British tunnellers began doing something very tricky: they continued digging tunnels at the same level as the Germans but secretly also put down shafts to a depth of more than 30 metres. This took them well below the German workings and the layers of watery, sandy, rocky earth, into the clay, which was much easier to tunnel through.

The Germans found themselves at a disadvantage when it came to mine warfare here because ironically, in holding the high ground, they were forced to dig a lot deeper to get to the relatively stable blue-clay layer. Higher up, even along the Wytschaete to Messines Ridge, the water table was close to the surface, and digging through the waterlogged liquid sand layer was almost impossible. Lower down the hill, the British shafts – especially with the use of metal shafts – could quickly penetrate the thinner and, in places, non-existent sand and the water table, and then hit the impervious blue-clay layer where real tunnelling could start.

The Germans sank concrete shafts using a circular steel 'shoe' with a cutting edge and, as it submerged under its own weight, they removed the earth from inside the shaft. Though clever, this was a process that needed a lot of heavy components: iron rings, sand, cement, metal rods, pumps, generators and a whole range of tools. Getting these to the front and set up would have been an extremely difficult process in peacetime, but in the mud and rain, across cratered trenches and churned-up earth, along narrow gauge railways and under fire, this was an amazing achievement.

The Germans expected the British to be facing similar problems, which lulled them into a false sense of security for a while. But when they picked up an insecure telephone conversation that alerted them to the presence of a big British mining operation at St Eloi, they called in their expert tunnellers and made aerial searches looking for the spoil dumps. These the British had been careful to camouflage, and the German experts reported that there was 'no immediate danger'. How wrong they were.[5] On 27 March at 4.15 am, the British blew six enormous mines, killing an estimated 300 Germans and destroying a large section of their frontline and support trenches. The centuries-old Mound was unrecognisable.

Even this turned out to be a wasted effort, though. The British infantry attacked only to be driven back, and the Germans now held high ground along the crater edge, a more defendable position that allowed them to observe the British lines.

Two important things came from the St Eloi attack. First, the British 172nd Tunnelling Company had for the first time successfully used a timber shaft to contain the layer of running sand, enabling them to reach dry clay at a depth of 12 metres. From there they began their sloping tunnel towards the German lines. This technique was to signal the beginning of the Allied dominance in the fighting underground on the Ypres salient.

Second, it had a destabilising effect on the Germans. They had always held the initiative underground and felt confident and dominant. Suddenly they realised that, unbeknown to them, the Allies had developed deep mining skills that would be difficult to counter. Fear and dread spread among the German engineers and tunnellers as they grasped that the initiative had shifted from them and they had been outfought and out-thought. In fact, after St Eloi the Germans would make every effort to countermine, but they would locate and destroy only one of the Allies' deep galleries (at Petite Douve in August 1916).

Even though mining operations and tunnelling were not proving effective and the casualties among the tunnelling companies were hitting 1000 per month, after St Eloi there was growing support within the High Command. The success of deep mining and the fact that for the first time it was the Germans who had seemed confused and beaten had given them new hope.

Like Griffiths before him, General Sir Herbert Plumer, soon to be given the overall command of the Messines attack, now saw value in offensive tunnelling and pushed GHQ to support it. But most importantly, the commander-in-chief, General Haig, was impressed, and so on 10 April 1916, he ordered that planning begin for the attack on the Messines Ridge. It was to begin, as Griffiths had suggested, with the firing of 20 massive mines. The countdown had started.

Haig believed the Allies should try to break through on the Ypres salient, iron out the bulging line along the Messines Ridge and push through to the coast. The French, under General Joffre, agreed that a massive push was needed to break through and end the war, but he had a different idea about where the push should be made. He wanted the attack to be nearly 100 kilometres to the south, along the River Somme, where pressure was mounting against the French. And so it was agreed that the Allied offensive on the Somme would start in July, in high summer, when clear skies and warm temperatures would make the attack less problematic.

Then came the news of the German assault on Verdun on 21 February 1916. A massive fortified salient that had epitomised French-frontier defence since Attila the Hun in the fifth century, it was the strategic gateway to the plains of Champagne and, beyond that, the city of Paris. While the British and French were sticking to the 'breakthrough' theory, the Germans had realised the futility of such an attempt. It had failed for them at Ypres, and they had watched as it failed for the British at Neuve Chapelle in March 1915 and for the French at the Battle of Artois in May 1915. Instead, they planned to draw the French armies into the narrow salient at Verdun, attack on three sides and bleed them dry.

As the Germans poured more than two million shells and sent 250,000 men against the French garrison, and as the French rushed men to Verdun, the British were asked to quickly shift men south from the Ypres area to restore the frontline along the Somme. The Australian forces, training and reorganising in Egypt after the evacuation of Gallipoli, were dispatched west across the Mediterranean to what would become the meat grinder.

A new phase in the Great War had begun.

NINE

Off to the Western Front

Oliver Woodward had little patience for parade-ground marching, yet there was one important drill that needed sorting the day before the embarkation parade. The officers had been issued with swords so that they could look the part of young and gallant warriors. They received only a few hours' practice in sword drill and saluting, certainly not enough to prepare themselves for this important parade. They had what Woodward called 'a very hazy idea as to what to do with our swords'. He cheekily added, '[This] was quite in keeping with our early education in matters military.'[1]

February in Sydney had been hot, and the late-afternoon southerlies that cooled the coastal suburbs had rarely made it out west to Casula, where dusty lines of tents seemed to stretch to infinity. Even the beer seemed warm, but the pub did a roaring trade as men made the most of their last local leave. The trudge back to camp from the pubs in Liverpool did little to dull a restless night on their thin straw palliasses, their minds on the big adventure that would sweep them from this comfort in the early dawn.

At 5 am, as the chorusing of kookaburras welcomed the new day, the damnable blast of reveille shook the camp. It was 19 February 1916. Men tumbled from their tents, tired and hung-over but excited and apprehensive about the day before them. There was much to do: pack their kitbags, clean their rifles, polish brass and prepare themselves for the final inspection. At noon the men paraded, were inspected by their commanding officer and moved in columns of three towards Liverpool Station. Cheered on by the crowd, their feet felt light as they stepped out, '*Hef-rite-hef-rite-hef-rite*,' to the thump of the bass drum of their unit band.

At 3 pm they boarded their steam train for the short journey to Sydney's Central Station. Forming up in Eddy Avenue, the men proudly kept in step as they marched up Wentworth Avenue, along College Street and to the gates of the Domain, where before a large crowd they were reviewed by the state commandant.

There, ramrod straight, their uniforms neat and their brass gleaming, was a company of the Shropshire Light Infantry. These British regulars could show the sloppy colonials a thing or two about drill. And how they did. In one swift movement, their salute with swords was conducted with a precision that took the mining boys' breath away, particularly after the pathetic sword salute by Woodward's mob, which he dubbed 'just an embarrassing mess . . . a sight to behold'.[2]

At 5.30 pm, the parade and farewell over, the men moved off to the Royal Agricultural Show Grounds where they were to bivouac for the night. This time, the cool zephyrs from Bondi made a difference, even keeping the mosquitos from Centennial Park at bay. But it was for most a restless night. Reveille was blown at 3 am, just as many finally began to sleep. In the breaking dawn they ate a light breakfast before parading for the final roll call. As each man's name was called, and after answering 'Sir', he stepped out to join a new line of men, those now checked off and going away. 'This last roll call was impressive; it seemed to mark the stage between playing at soldiers and being soldiers,' Woodward reflected.[3]

The sun was just breaking the eastern sky as the men moved off for the short march to Woolloomooloo Bay and the waiting troopship HMAT *Ulysses*, designated during the war as the A38. There were crowds in the streets, many bidding their last farewells. Some joined their loved ones in the ranks, swinging along in step. It was a sad sight. Oliver Woodward was thankful there were no relatives seeing him off, for he found it heartbreaking to witness those last painful moments between mothers and sons. 'I confess that had my own mother been present I would have broken down,' he wrote.[4]

At 8.30 am the ropes fell away and the last of the coloured streamers broke and settled in the murky harbour water. Slowly the water churned, and the A38 moved out into the harbour, passing a number of launches bearing crowds of waving Sydneysiders. An hour later the *Ulysses* had cleared the Heads and left the last of the launches bobbing in its wake. Then the sandstone cliffs and eucalypt horizon slowly faded from view.

On all departing troopships this moment was difficult for every man. The deep longing to be back with kith and kin, and the fear of what lay ahead, was everywhere evident on the ship. Woodward cast his eye about the faces lining the rail, outwardly proud and enjoying the great adventure, but inwardly sullen and apprehensive. Years later when he looked back on his war experiences he would note: 'One passed through many dark days at the Front, but there one was buoyed up with the thrill of battle and nothing seemed quite so bad as the day when steaming down the coast of New South Wales, one had to battle with the fear of the unknown without being keyed up as in War.'[5]

For many men, this was their first time at sea, and seasickness turned many into 'a sorry spectacle', irrespective of rank. For two days the *Ulysses* journeyed south, arriving in Port Melbourne, where the men were disembarked and taken to Broadmeadows Camp. After a week of parades, march-pasts and presentations, they again boarded their troopship and headed west across the Bight to Fremantle. They enjoyed two days' leave in Perth, having a dip in the surf and a lunch prepared by the women of Fremantle, then boarded their ship, along with 800 reinforcements. The ship departed for their anchorage in the Gage Roads, in Fremantle Harbour, but struck an uncharted rock and started to take water, so the troops were quickly offloaded and marched to Blackboy Camp, about 20 kilometres northeast of the city of Perth. It was three weeks before the ship was repaired and they were able to sail to war.

On 1 April 1916, transport A38 moved out from Fremantle Pier to the blessing of the Anglican Bishop of Perth and the cheers of the throng of people perched on every available vantage point. Slowly the ship headed into the setting sun and safely anchored in the Gage Roads. At six o'clock the following morning, they set sail westward, the men lining the rails for their last glimpse of Australia: the small, sandy dollop that was Rottnest Island.

Aboard the *Ulysses* there was an added level of anxiety and tension, because on 1 March the Germans had announced that they were extending their submarine campaign. Australia was far from the U-boat ports of northern Europe, but the operational capability of these submarines was not known, and their success had put the fear of a watery grave into many travelling at sea for the first time. There was also the potential for German surface raiders to appear such as the *Emden*, which was sunk in November 1914 on this same convoy route.

Before the war, the *Ulysses* was a comfortable passenger ship of the Blue Funnel Line, but now 2200 men were crammed aboard and her cargo decks had been converted into sleeping quarters and mess decks. The men slung hammocks in rows from the roof, and at night their curved backs provided an unusual image, arcing into the darkness and offering the officer on duty a strange sight as he completed his nightly rounds. That and the sounds of sleep; the snoring and the music of the night of so many men, reminded Oliver Woodward of 'an invisible choir'.[6]

While the men swung in the blackness of the hold, the officers fared a little better. Second lieutenants – the lowest-ranked officers – were four to a cabin, lieutenants were three to a cabin and captains two. Those ranked major and above had a cabin to themselves. Officers also had access to the ship's dining room, unlike the men, who sat at long mess tables.

The days were filled with training, lectures and boredom. There were lessons in the care and maintenance of equipment, rifle cleaning and basic field tactics. Men exercised in the narrow confines of the deck or watched the flying fish launching themselves from passing waves. At night, the portholes were closed and the ship blacked out. Men had little to do but write up their diaries, play cards and two-up, and get to know their new battalion friends.

As on all Australian troopships, the crossing of the Equator was a big day. From 2.30 pm on 12 April, the *Ulysses*' crew and the officers devoted the afternoon to a fancy-dress King Neptune ceremony, and rough-and-tumble games for the men. As it was a dry ship, the fun was contained and the men soon slipped below for dinner and an early night.

The first sight of land after leaving Australia was Cape Guarda-fui, on the northeasterly tip of Somalia. Then the ship made a tight turn to the left and headed into the Gulf of Aden. Passing French Somaliland and the port of Djibouti, the *Ulysses* went through the Strait of Babel Mandeb and entered the Red Sea, where it was met by three British warships, to the relief and pride of many of the crew. Travelling north past the island of Perim, on through the oppressive heat of the Gulf, the ship finally made landfall at the southern entrance of the Suez Canal, where they anchored to await orders.

On 23 April, they entered the Suez Canal, which was protected by British soldiers and artillery. After passing through the Bitter Lakes, as night fell they came to a section of the defence line held by Australians. From the canal bank Australians yelled out, asking who they were and where they were heading. At one point, a voice from the dark bank asked, 'Is Jim Binny on board?' and soon Sergeant Binny was exchanging news with his brother.[7]

Just before dawn, the *Ulysses* reached the northern end of the Suez Canal and dropped anchor at Port Said. Alongside was a hospital ship, painted white with a large green band along the hull and distinctive Red Crosses on the sides and deck. At night, the ship was illuminated by red, green and white lights creating, to Woodward's eyes, a beautiful picture. During the day, the Australians watched as a 4.7-inch gun was mounted on the *Ulysses* and a naval gun crew taken aboard. This and the Italian cruiser that appeared as an escort focused the men's minds on the increased danger of German submarines. Creeping into the darkness of the Mediterranean in a blacked-out ship, all were happy to accept the discomfort of continually wearing life jackets.

Their journey was cut short when the *Ulysses* was deemed unsafe, despite being repaired in Perth. They were ordered into the Egyptian port of Alexandria, at the mouth of the Nile. The men of the various battalion reinforcements disembarked to join their battalions camped at Tel el Kebir, and the men of the mining battalions were transferred to HM Transport B.1., the Cunard liner *Ansonia*. Once on board they were told there would be no shore leave, which was unpopular news after they had been cramped on the *Ulysses* for more than three weeks. And being the men they were, 117 of them ignored the order and went over the side, setting off for town. But the provost marshal, the head of the British military police, was one step ahead of them. Not only were they rowdy, undisciplined Australians – and he knew their type well – they were miners. As the Australians walked from the ship towards the tram terminus, they passed along a straight stretch of wooded road. And what should come along conveniently but some empty trucks? The Australians flagged down the drivers, who invited them aboard to save them the walk. But they never made it to the tram terminus. The provost marshal had simply collected all 117 men. Instead of an evening in town, they spent an uncomfortable night under the guard of the local military police, who threatened to send them home to Australia. The men were pleased the next day to see an armed detachment from the ship arrive to escort them back, and despite having been tricked they gave the provost marshal a hearty farewell.

The following morning, Oliver Woodward and his mates from the mining companies, along with British soldiers returning to England, left Alexandria and headed west. Once at sea, they were joined by a British destroyer, which established a zigzag course for the ship and at intervals would completely circle the ship at high speed to ward off submarines. Three days later, they began their slow entry into the Maltese port of Valletta, a dangerous exercise after the sinking of a British destroyer and two mine-clearing trawlers the day before. Mines had been laid by a rogue vessel flying the Greek flag, and 120 mines had been located by mine-sweepers and destroyed.

That night, the men were allowed ashore to stretch their legs in the historic port. This tiny rock at the crossroads of the Mediter-ranean had seen many warriors arrive on its shores, from the Phoenicians, Romans and Saracens, to the Normans and Knights Hospitaller and, more recently, the armies of Napoleon and Nelson. The arrival of some Australian tunnellers was perhaps a non-event, but for many of the Australians it was their first expe-rience ashore and a wondrous marvel to their colonial eyes.

Their four hours' shore leave gave them the chance to have a feed and knock back a carafe or two of the local *vin rouge*, some-thing they had probably not previously experienced but would come to know well over the next few years. For dinner, most passed on the offer of *Fenek Moqli*, or fried rabbit, as they ate 'underground mutton' so regularly at home, but they did try the thick vegetable soup, the spaghetti and octopus, and the baked macaroni. Woodward, rostered as the officer in charge of the guard, viewed the city from the deck of the *Ansonia* and could enjoy only the smell of cooked garlic and freshly baked Maltese bread.

The coast of France broke the horizon two days after leaving Malta. Running down the coast, they sighted the lights of Marseilles and by 8 pm had dropped anchor in the outer harbour. The next day, after berthing at the quay, the men packed up and prepared for disembarkation. It was just on dark as they bade farewell to their transport and marched up the long hill from the quay to the Paris Lyon Railway Station. Lining the footpath was a polyglot of faces reflecting the city's long history. For the Australians these were strange faces: Corsicans and Armenians,

Greeks and Italians, Spaniards and Berbers and people from North Africa. They wore strange clothes and spoke a mix of languages of uncertain origin.

After climbing the long stairs to the station, the men found themselves confronted by a strange sight. A long trestle table was erected and a quartermaster sergeant was handing each man a sheepskin vest and two blankets as he passed along. At the end of the platform, groups of men had gathered and, dressed in the new vests made from the finest Australian merino, stood around bleating and laughing. The French railway staff were perplexed at the sound of so many soldiers *baaing* and must have wondered who had been sent from far across the world to fight their war.

Amid the cacophony, Oliver Woodward stood and stared at the giant engine. The locomotive was the largest he'd ever seen, to say nothing of the long line of carriages and rolling stock that completely filled this equally long platform. At 11 pm the men boarded the train, made themselves comfortable and prepared for the journey. Half an hour later, steam blasting across the vacant platform, the train and the long trail of carriages slowly began to move forward.

The dawn light revealed the beauty of the Rhone valley, 'a perfect natural garden' as Oliver Woodward observed.[8] A stop for breakfast at Orange and the train pushed on, skirting the Rhone River as it headed towards Valence and Lyon. Unlike the dry scrub and rough paddocks of Australia, here the land was heavily tilled with row upon row of crops, without fences to break the extended beauty. There were hedgerows, green with the fresh growth of spring, and line upon line of fruit trees resplendent with blossom. Scattered about were villages of red-tiled stone houses, their long, thin, shuttered windows and ornate doorways like nothing the

99

men had seen before. Wisteria and climbing roses clung to the walls and pencil pines reached for the sky.

Late in the afternoon, the train slowed and came to a stop at Lyon. The station was overrun by the French Red Cross Auxiliary, which scurried about serving the men sausage and bread. Little did the Australians know they were in the gastronomic heartland of France, sandwiched between the great wine-growing regions of Beaujolais and Côtes du Rhône. As hunger took over, they got stuck into their *saucisson de Lyon* and *baguette*.

As the train sped north towards Paris and their destination, Hazebrouck, the reality of war began to sink in. Woodward wrote:

> The charm and glory of these peaceful surroundings would suddenly seem dispelled with the realisation that we were soldiers speeding forward for active warfare, the great unknown ahead of us. We seemed to envy the inhabitants of these peaceful villages, and then the realisation would come that they already knew the horrors of War – the fathers and sons already fighting for their country. We journeyed through a land seemingly occupied solely by old men, women and children.[9]

The colonel believed that the fact they were on a train should not in any way disrupt the appropriate daily operation of a military unit. An orderly room was established, and orders were sent. These were yelled out of carriage windows, passing by word of mouth along the length of the train. In typical fashion, the men took great delight in adding to or altering these commands, as in Chinese whispers, so that confusion and chaos were the real order of the day. 'It was a polite reminder that all hands wished to be left in peace,' wrote Woodward.[10]

The train passed south of Versailles. Through the flickering breaks in the trees that lined the track, the men could see the Eiffel Tower before they sped on: Rouen, Boulogne and Calais for breakfast. Then the train turned east. The sky was now heavy with cloud, and the rain, grey and cold, cast a monochrome wash across the countryside, so very different from the beauty and sunlight in the south.

The paraphernalia of war began to appear. Near Paris they saw a military airfield with lines of aircraft. And now there seemed one endless military camp: tents, wagons, vehicles of all kinds, lines and lines of huts, horses, artillery and stores areas. At 11 am on 8 May 1916, Oliver Woodward and the rest of the Mining Battalion finally climbed down off the train, their legs stiff and their heads heavy and tired. After unloading their stores, the officers were allocated billets in town with French families, while the men were sent to a large military billet not far from the station. As Woodward settled down in another strange bed and tried to make himself comfortable on one of the odd sausage-shaped pillows so popular with the French, he could hear the far-off rumble of the guns from the front. He noted in his diary: '[I] began to realise that very soon we would receive our baptism of fire.'[11]

TEN

Back Near Messines

Though the winter of 1915–1916 had slowed down offensive actions on the surface, the war underground had continued apace. From the high ground at Hill 60 and the Caterpillar, south through St Eloi and Wytschaete, to the end of the Messines Ridge, the struggle raged on, not only against the enemy but against the cold, rising water, the wet, flowing sand, and layers of pebble and clay. Tunnelling was never allowed to stop.

Within just a year of Griffiths suggesting the formation of specialist tunnelling units, there were 27 tunnelling companies, comprising 20 from Britain, three from Australia, three from Canada and one from New Zealand. Each company was made up of roughly 1000 men. Twenty-thousand men represented a whole division: a lot of men to take out of the frontline, especially given their questionable contribution to the war effort, the vast quantities of material they required and the amount of rum they seemed to consume.

Griffiths, under pressure for the tunnellers to perform, had another brainwave: mechanical boring machines were used in coalmines and in the clay below the streets of London, boring the underground railway. If such machines were put to work on the Western Front they could speed up tunnel construction and maybe even save lives. With his usual enthusiasm and passion, Griffiths approached General Harvey with the idea, suggesting that the army order six machines, have them fitted with the appropriate cutting blades for clay, and get them to the front as soon as possible. He was only able to convince Harvey to bring one machine, and headed to England to speed its dispatch.

The Stanley Heading Machine, manufactured in Nuneaton, Warwickshire, was assembled underground at Petit Bois just west of Wytschaete, between Messines and St Eloi, and set to work on 4 March 1916. At first it worked wonderfully well, producing a tunnel two metres in diameter – considered large at the time – at a rate of one metre every two hours. Yet it tended to dive rather than move forward horizontally, and even the expert who had been brought from England to assemble and operate the machine could not control it. When it was stopped for service and routine mainte-nance after several hours, it was difficult to restart. By the time adjustments were made and the borer started up again, the pressure of the clay was such that the machine was wedged in the tunnel and could not move forward. General Harvey, to say nothing of the tunnellers, lost faith in the Stanley Heading Machine and, after trav-elling only 60 metres, it was abandoned. Today it still sits 25 metres below the surface, locked into the blue clay that claimed it in 1916.

Griffiths had lobbied heavily for this machinery, staking his reputation on its potential. After this mishap, although he was still full of energy and enthusiasm, he was effectively sidelined. The tunnelling companies were now far more organised and profes-sional than when he helped create them, and the new layers of command sadly did not include him. He requested two months' leave for personal reasons and, after a final visit to the front, handed over his beloved and trusty Rolls-Royce to Harvey. On 30 March 1916, his work done, he took a steamer to England, never to return to the front again.

Instead, Griffiths settled back into family life with his wife, Gwladys, and their four children. He resumed his duties as an MP and worked at the Ministry of Munitions until, on 4 November 1916, he was ordered to report to the Department of Military Intelligence. There he was assigned a secret mission that was akin to something out of a *Boy's Own Annual*: to go to Romania and destroy the oilwells, oil, petroleum and grain stockpiles before they fell into the hands of the advancing Germans.

With a number of expatriate Brits, Griffiths drained storage facilities of oil and petroleum, and either burnt them or drained them into holding tanks to permeate into the soil. He destroyed derricks; plugged wells with anything at hand; smashed lathes, electric motors and any tools with sledgehammers; and destroyed grain stocks.

With the Germans often just hours behind him, he raced from one facility to the next. Many times, he had just left a burning, mangled oil facility as the Germans arrived; his car streaking away into the distance. The Germans were closing on him all the time. Twice their patrols cut him off and it was only through the speed of his car outdistancing the German cavalry that he escaped with his life.

At Moreni, which held the most valuable oilwells, he became trapped in his own fire and, had it not been for the bravery and quick thinking of Captain Bibesco, a Romanian aristocrat and later the son-in-law of Prime Minister Asquith, Empire Jack would have been lost in the flames.

It was estimated that Griffiths caused £55 million in damage at 1917 prices.[1]

In late February 1917, Griffiths smuggled himself over the Russian frontier and took the train to Petrograd. There he found the revolution in full flight, demonstrations in the streets, food shortages and the army now joining the workers. Finding the British Embassy and reporting on his work, he was issued with a new passport and documents to see him safely across the border.

But from Britain came Jack's last orders: to meet the Tsar and report back to the British Foreign Office. Some days later, Griffiths was ushered into the presence of Nicholas II and was invested with the Order of St Vladimir, an honour to go with the Grand Star of Romania that he had received from King Ferdinand before his departure from Romania. This meeting was the final attempt by Nicholas to persuade his cousin, the King of England, to provide him and his family with asylum in Britain as the revolution swelled about the royal palace. It was only two weeks later, in mid-March 1917, that Nicholas abdicated. Empire Jack was probably the last person to be decorated by the Tsar.

By Easter Griffiths had returned to his family, and to grateful thanks for his work. He was made Knight Commander of the Bath by the British, and the Légion d'Honneur by the French for his service and deeds. It was a fitting end to three years of war. His company, Griffiths and Co., remained in a holding pattern, carrying out small projects and military contracts to keep the business ticking over. In 1917, he added Norton to his name by deed poll, to become John Norton-Griffiths.

After the war he formed the Comrades of the Great War Association, which later joined with similar organisations to form the Royal British Legion, the UK's equivalent of the RSL. In 1922 he was made Baronet of Wonham. He remained a Member of Parliament until 1924.

Norton-Griffiths' next major project, however, would be his final undoing. Tenders were being called for the raising of the wall of the Aswan Dam in Egypt. Norton-Griffiths' tender was submitted in late September 1929, just four weeks before the New York stock-market crash. He won the contract, but soon financial problems plus continual clashes with the Egyptian authorities slowed the work, and finally it stalled. These problems affected his health. After writing to the Egyptian government in a last desperate attempt to salvage the project and get new financing in place, he travelled north to Alexandria, where at San Stefano he was seen to take a small rowing boat out into the bay. Sometime later his body was found floating with a bullet hole in the head. There were suggestions that he had been murdered by the Romanians because of his earlier destruction of their country, but the general consensus was that this was suicide. He was just 59.

So ended the life of an amazing man, an independent, courageous free thinker and doer who had a vision both for himself and the country he served.

✕

In the summer of 1915, the Royal Engineers of the 175th Tunnelling Company had started what would become a very important strategic tunnel at Hill 60. Beginning 200 metres behind the British frontline, they had driven a long horizontal tunnel in the direction of Hill 60. Approaching the railway embankment, they began sloping the tunnel downward, but not without great difficulty. First they struck the shallow water table, then below that, watery, sandy clay that proved almost impossible to dig through and to retain in any safe and stable way. The shoring broke and the liquid earth seeped through the cracks, spurting onto men's faces and clogging up their pumps. To counter this, a wooden caisson was introduced, which created an airtight chamber for the men to work in.

And their efforts paid off. Deeper, they hit the blue clay, and with the efficient clay kickers on hand, quickly pushed a narrow gallery towards Hill 60. Nearly 30 metres deep, it went under the German lines and became known as the Berlin Tunnel because, the men joked, at this rate they would soon be under Berlin. It would become the service tunnel for the mines on Hill 60 and the Caterpillar, and be fought over, flooded, attacked, collapsed and countermined for the next 18 months. They also continued shallow mining, just five metres below the surface, to divert German attention from the deep tunnelling.

During fighting in early 1916 at the nearby Bluff, a German miner was captured and told of a German gallery that was being built to blow up the bridge across the railway cutting, which was on the British frontline. The British started a new countermining operation. Working fast, the tunnellers dug a branch gallery and fired a number of mines, collapsing about 60 metres of German tunnel and wrecking the enemy's entire system. But the explosion also seriously damaged 60 metres of the Berlin Tunnel, an unfortunate side effect. A German tunneller later captured on the Somme said that the blast had travelled along the German tunnel and destroyed a shaft near the surface, and that the ground was so shattered after this explosion that the Germans were never able to sink a shaft in that area again.[2]

In April 1916, the 3rd Canadian Tunnelling Company relieved the exhausted Royal Engineers. They entered the shattered Berlin Tunnel and found that 250 metres were still intact. After clearing the debris, they began driving new galleries under Hill 60 and the Caterpillar. Within three months, they had rebuilt the shattered system and dug a tunnel well beneath the German tunnelling system, and well under the enemy frontline. The Canadians then began a clever defensive system by digging a series of intermediate tunnels between their deep mines and the shallow system above.

Beneath, the German miners sweated in the dark passages, breathing in the foul air and fearing for their lives. Just like the British miners, the men worked in silence, hunched over, fighting the oozing, liquid sand that extended down four or five metres in this area, or working the hard clay with a *sticheisen* – a tool like a large apple corer – which was a far less efficient way of mining clay than the British 'clay kicking' method. Then they filled their sandbags, slowly and carefully, ensuring they avoided each other in the low light of the narrow gallery, and dragged them out.

In silence, the guts of Hill 60 were slowly torn out. Bag upon bag was filled, tied off and hauled down the long, timber-lined galleries. In the early days of mining the Hill there were no electric mining lamps, and miners were loath to use carbide and further foul the already putrid air. And the problem was always air. Although they had installed mechanical ventilation systems of various kinds, which had improved the air quality and their comfort, there was always the associated noise, whether it be the hum of the motor or the huffing of the air.

Since the Allies had lost the ridge line, they had been at a continual disadvantage. The Germans consolidated their defences, built concrete blockhouses and underground shelters, and dispersed their artillery in well-constructed positions on the ridge to fire upon the Allied front. Griffiths' original plan was for just six deep mines, but once a major mining offensive was agreed upon, this was expanded to many more targets along the salient. Now the plan was to prepare 23 mines at depths of 30 metres, not the

15 metres originally planned, and to double the amount of explosives. Secrecy was of primary importance, and the mines were always referred to as 'deep wells in connection with water supply', a ruse that seemed to work.[3]

The northernmost mines were under Hill 60 and the Caterpillar. The next mine south down the line was at St Eloi. The large British blows in March 1916 had stirred up a hornets' nest in this area and the Germans, after re-taking the craters and their lost ground, kept up a relentless artillery and mortar bombardment of the British lines opposite.

In the spring of 1916, the 172nd Company was replaced by the Canadians, who immediately commenced a deep shaft well back from the frontline. Within three weeks, they were down to 30 metres, had put in a station ten metres down and had driven a gallery to contain a power plant and other equipment. Here the Canadians worked in very difficult conditions. The soil was wet and fluid, the hard clay was unstable and cave-ins were common. In other places the ground was shattered and fragile and old galleries criss-crossed the area, making tunnelling dangerous. At one point, the main gallery flooded, stopping work, but the Canadians soon had the water pumped out, the silt removed and the tunnel operational again. They then pushed a gallery forward, deceiving the Germans and outflanking their countermining efforts, finally driving under the German frontline and on below the old craters from the 1916 explosions.[4] They enlarged a gallery well behind the German frontline, and by late May 1917, they had completed the placement of a massive charge of 50 tonnes of ammonal, the biggest in the long history of military mining.

Along the next 900 metres of the frontline, tunnelling was next to impossible in the sandy clay, so the defensive systems were only five metres deep, which made them vulnerable to cave-ins through shelling. When the Canadians took over this section of the front in 1915 from the 250th Tunnelling Company of the Royal Engineers, they immediately began deep mining on the Hollandscheschuur salient. They dropped a shaft to 18 metres then pushed out a gallery that was quickly driven under the German lines 40 metres away, and then to a point 100 metres beyond it. Fearing a German trench raid would reveal the opening to their shaft, they carefully concealed the opening by making it resemble a dugout protected by a machine gun between two inclined entrances. They also built a second shaft to ensure access and an escape route. The Germans were well aware of the Canadian tunnels and fired a large charge in May 1915 that created a crater they called *Cöln*. Just as the Canadians had finished their construction and were about to lay their charges, the Germans exploded another large charge in June 1916, creating a crater they named *Cassel*. This slowed the Canadians' work, but the first of the three Hollandscheschuur mines was ready by 20 June 1916, the second mine by 11 July and the third mine by 20 August, just under a year before they would be needed.

A little further south at Petit Bois, tunnelling was commenced in late 1915 by the 250th Tunnelling Company of the Royal Engineers. They drove a shaft to a depth of 18 metres and pushed out a gallery, before deepening the shaft to 30 metres. By early June 1916, the British had dug a 450-metre-long tunnel under and past the German frontline. However, on 10 June, at 6.30 am, the Germans blew two heavy mines that exploded above the British tunnel, causing 75 metres of gallery to collapse and trapping 12 tunnellers in an undamaged section near the face, where the tunnel was only one metre square.

As in mining communities across Britain, the tradition among tunnellers was to always attempt a rescue, no matter how slim the chance that anyone had survived. A rescue team raced to the shaft and with Proto breathing equipment they were soon underground at the caved-in area of tunnel. Clearing the long section of collapsed earth and shattered timbers was very slow. A Welsh rescue specialist named Haydn Rees decided instead to dig around the collapsed part and build a parallel tunnel. Frantically they dug, oblivious to the noise they were making and desperate to get to their entombed mates.

Though the explosion had shattered the 12 miners' nerves, they were all uninjured. Finding they were trapped and only a tiny amount of air was entering through a small ventilation pipe, they began to tear away at the earth, but soon an argument broke out: some believed they should dig themselves out while one experienced miner from Cumberland, Sapper Bedson, argued that they should stay calm, lie still and conserve the little air they had.

Desperation took over, and a deep animal instinct to survive. The men ignored Bedson's suggestion and attacked the face, dragging away broken timber and clawing at the sticky clay. But soon the exertion left them heaving and convulsing on the wet tunnel floor, gasping for air that was now heavy and foul. And then the tiny trickle of air that had sustained them suddenly stopped, and a new panic overtook them.

It was now late in the afternoon. Bedson suggested to the men that they spread out along the length of the tunnel and rest, breathing slowly to conserve the air. He crawled to the face, made a bed of sandbags and, after taking the glass from his watch so he could feel the time, curled up and drifted into a light sleep. His end of the tunnel was slightly higher than at the site of the collapsed face and, as a result, the air was a little better. He felt some comfort in the knowledge a party of coalminers that had been trapped for 13 days was rescued safely. If he could just hold on, a rescue party would get him out.

Early the following morning, the men started to spread themselves along the tunnel, but at 5 pm the first man died. The others quickly followed and by the end of the third day, 11 men were dead, spread along the tunnel. Only Bedson was alive. Though he was ravenously hungry, he did not eat the two hard army biscuits he had in his pocket for fear these would make him thirsty. He also had a water bottle, but would only sparingly wash out his mouth and return the water to the bottle. He kept his head, kept himself warm and kept track of the time.

Meanwhile, Haydn Rees and his tunnellers were desperately working forward. Instead of the usual rate of four to five metres a day, now they were clearing out more than 12 metres a day. But they had a long way to go: a daunting 75 metres and straight towards the Germans. By the fourth day, they still found the tunnel beside them broken and collapsed. There would be little hope of anyone being alive after this time, and the order went up to the surface to prepare 12 graves. Yet they pushed on.

And then, six-and-a-half days after the Germans blew their mine, the rescue party broke into the stinking tunnel. In the narrow shaft of the officer's torch beam, the line of bodies could be seen extending towards the face. The rescue party withdrew to the surface and reported the sad loss of the tunnellers lying below.

Crouched at the face on his bed of sandbags, Bedson had neither heard nor seen the men of the rescue party, but he did notice that the air felt a little fresher and the pressure had dropped. He crawled along the tunnel and over the bodies of his dead comrades, until he came unexpectedly to the hole made by the rescue party. Just as he did, the rescue team returned and was astonished to see sapper Bedson still alive. Exhausted, he extended his hand, saying, 'It's been a long shift. For God's sake give me a drink.'[5]

The rescue party carried Bedson back along the new tunnel and out into the fresh air. He was given more water then loaded onto a stretcher and carried along a communication trench towards the rear and the Casualty Clearing Station. Suddenly, the ground erupted with shellfire – he was not out of harm's way yet. Fortunately he arrived and was inspected by a medical officer, who wrote that he was 'clear and rational' and basically unhurt from his long and frightening ordeal.[6] He was rushed back to England.

Bedson had already been wounded on the Ypres salient in 1914, had recovered and then been sent to Gallipoli. There he was wounded again, recovered and again sent to the trenches of Flanders, though this time as a tunneller. Even after his latest awful experience, Bedson requested to return to his tunnelling company and resume work, but it was felt he had already contributed enough to the war, so he was given a job at a Base Depot safe behind the lines.

The tunnellers at Petit Bois eventually drove a shaft nearly 600 metres long, and at the end dug two separate tunnels splitting off in a Y shape, which they then charged. The Germans countermined, blowing three charges from the large mine craters above. These did not destroy the British mines, but believing they had, the Germans suddenly ceased their work, leaving the British to quietly maintain their mines.[7]

Just to the south of Petit Bois was the Maedelstede Farm mine. Here two shafts had been built and two tunnels pushed forward with the objective of Wytschaete Wood some 800 metres away, well behind the German lines. Light railways were constructed that worked hard at night, shipping spoil back well behind the line and out of sight of the Germans, and bringing up food, ammunition and materials.

Next, heading south, was the mine at Peckham Farm. Started in late 1915, it progressed slowly because many of the men fell sick due to the severe winter. By early January, the tunnellers were down to a depth of 20 metres, where they began work on a tunnel towards the German strongpoint. The ground was heavy and the clay, which had a high moisture level, swelled and snapped the lining timbers of the tunnel like celery. Heavier timber had to be brought in, which also slowed the work.

By March 1916, the Canadians had taken over and the tunnel had extended to 150 metres. The Germans began heavily shelling the shaft workings and entrances and raided the trenches. In late April, the tunnellers were called upon to line the parapet and fight off an attack, and the heavy German trench mortars known as *minenwerfers* forced the closure of three shafts.[8]

By the end of April 1916, the Peckham chamber had been charged with 35 tonnes of ammonal, but a series of collapses and flooding at one point cut off access to the charges. The tunnellers dug diversion galleries, but these, too, struck bad ground. In the end, a parallel tunnel three metres above the old tunnel was dug, and the Canadians were able to break back into the old shaft 300 metres out, just ten weeks before the mine was to be blown.

The 250th Company had also begun work in December 1915 on the Spanbroekmolen mine, 400 metres south of Peckham Farm. They succeeded in getting a shaft down through wet ground that was difficult to mine, then handed over to the Canadians in January 1916. The Canadians in turn handed over to the 171st Company and by June 1916 the tunnel had been completed and the charges laid. With time on their hands, the ambitious tunnellers started a tunnel towards 'Rag Point', 360 metres further behind the German lines. The tunnel had gone 350 metres by mid-February 1917, when the Germans blew a camouflet. Then, just as the tunnel was passing under the German frontline at Narrow Trench, another German camouflet collapsed 150 metres of trench. Not wishing to draw German attention to their deep shafts, the Rag Point objective was abandoned.[9]

Work started at Kruisstraat, 600 metres south of Spanbroek-molen, in December 1915 on what would become the longest tunnel under Messines Ridge. It stretched 700 metres from the shaft off the Kingsway communications trench, to well behind the German frontline. Four separate explosive charges were laid, but the men continually battled the problem of water seeping in. Special sumps and a major bailing effort were required to maintain the charge, and work was completed less than a month before zero hour.

South of Kruisstraat, the tunnellers found a very different soil profile to the surrounding area. Waterlogged sand went down to a depth of 30 metres, making it almost impossible to build a stable shaft. Three were attempted, and the last one, dug in February 1917 at Boyle's Farm, was deep enough to allow them to drive a gallery and begin offensive tunnelling. After pushing forward, they dug down further, to a depth of 35 metres, where they struck the blue clay. A new drive forward began, 35 metres below no-man's-land. Suddenly, disaster. At 165 metres in, the tunnellers working at the face hit bad ground, and water and quicksand poured in. They retreated and built a strong dam, but not before 30 metres of tunnel had been lost. Above, German shelling smashed the shaft head and left several men dead. Doggedly, the tunnellers returned to the face, pushing their shaft forward and under the German strongpoint, Ontario Farm, about 1.5 kilo-metres west of Messines village.

On 24 August 1916, in a deep gallery beneath Petite Douve Farm, the British unexpectedly broke into a well-built German gallery, making a large and very obvious hole in the enemy tunnel. Quickly the British tunnellers blew a camouflet, hoping to smash the German gallery and flood their workings. It did little damage and the Germans fired a heavy camouflet on 27 August, completely smashing the British gallery, so much so that the workings were rendered useless and were immediately abandoned. This was a great disappointment for the Canadians when they took over, but they had the last laugh. To deny the Germans access to the destroyed Allied mine workings, they ceased pumping out water, allowing the tunnels and galleries to flood. Because of the porous nature of the soil and the shattered ground, water easily found its way into the German workings, flooding them, too. The Germans were diverted into pumping out their own tunnels. And the Canadians kept them pumping: they dammed water and released some every time the water level dropped in the German workings.

Below the southern end of the Messines Ridge, just north of Ploegsteert Wood, the British 171st Company started a deep shaft and tunnel from their frontline at Trench 127 in December 1915. They dropped a shaft to a depth of 25 metres and pushed towards the German frontline at what was known to the British as Ultimo Switch. But at the 215-metre mark, without the slightest warning, quicksand burst through the face. The tunnellers fled back down the gallery as the liquid sand swirled around their feet, and quickly built a dam. Major Edgeworth David believed the tunnellers had accidentally encountered part of the ancient pre-glacial bed of the Douve River.[10] To avoid the bad ground they sank a new shaft to a depth of 27 metres, dug a side gallery off to the left and charged it with 16 tonnes of ammonal. The main drive went forward another 415 metres and was charged with 23 tonnes of ammonal. And so by mid-May 1916, another two mines lay silent and ready for the big day.

The last and most southerly site was a twin mine at Factory Farm, 300 metres south of Trench 127. Starting in February 1916, the tunnellers dug a shaft down to a depth of 25 metres then pushed a tunnel out, forking at the end to form a Y. They laid a charge of nine tonnes of ammonal in the northern branch and a further 18 tonnes in the southern branch, along the German frontline.

Four other mines were excavated and charged at a site to the east of Ploegsteert Wood known to the British as the 'Birdcage' and to the Germans as the 'Duckbill'. South of the extent of the June 1917 attack, they were not blown but kept in reserve for a later time.

By early 1917, a necklace of enormous mines was being created along nine kilometres of German frontline. And the Germans were all but unaware of what lay beneath their feet.

ELEVEN

Just out from Armentières

When the Australian Mining Battalion arrived at Hazebrouck, they were informed by the controller of mines of the 2nd Army that their present formation as a battalion did not fit the current British operational formation, and so the three tunnelling companies became separate entities. This break-up of the battalion pleased the independently spirited men, as they preferred to be part of a smaller unit that operated independently. They sewed their own distinctive colour patches onto their uniforms: a purple 'T' with the numbers, 1, 2 or 3 designating their company.[1] Three senior officers formed a separate new company known as the Australian Electrical and Mechanical Mining and Boring Company, which, because of its long and silly title, became known to the men simply as 'The Alphabetical Company'. It established a large workshop in the town of Hazebrouck, and provided support for all the Allied tunnelling companies for the remainder of the war.

The three independent Australian tunnelling companies were now dispatched to various frontline areas of operation around Armentières. The 1st Australian Tunnelling Company was to operate north of Armentières and have their headquarters in the town. The 2nd was to operate south of Armentières, with their headquarters in the village of Sailly-sur-la-Lys, and the 3rd would operate near Lens, with its headquarters at Nœux-les-Mines.[2]

To give the men practical skills and familiarise them with frontline conditions, each company's four sections were attached to experienced active units such as the Royal Engineers or the Canadians. There were two distinct types of ground in the vicinity of the British frontline – hard chalk around Lens, which was in the middle of the British front, and blue clay along the Messines Ridge. As each required different mining skills and equipment, once a company was located in an area and had come to know the ground, the mine layout and the enemy workings, it was rare that these units were moved. If it was imperative they transfer, it was done over a period of up to two months to give the relieving tunnelling unit time to become well acquainted with the new position. This was particularly the case with offensive mining operations, as this work was very delicate and highly secretive.

Six days after their arrival at Hazebrouck, the company commanders, adjutants and officers in charge of sections, including Woodward, were taken on an inspection of the areas of the frontline where they were to begin work. Here they were shown mines, underground workings and maps of tunnels. Next the men were lined up in front of the quartermaster and each was issued with a rifle, ammunition and basic webbing.

On the Sunday before the men of Woodward's tunnelling company moved out to their operational area north of Armentières, they attended a church parade in the local town hall. The men of the AIF were not especially religious and would generally avoid church whenever possible, but given they were about to go into the line for the first time, perhaps thoughts of a Higher Being crossed their minds. It was something of a serious and sombre occasion, but partway into the service, loud yells and screams came from the hall immediately above. Later it was discovered that the locals were conducting a cockfight to determine the district champion. Even given their brief piety, had the men known, said Woodward, 'they would have been sorely tempted to depart en masse from the service and try their luck in picking a winner'.[3]

On 15 May 1916, the men of the 1st Australian Tunnelling Company boarded old London buses, the classic red finish replaced by drab grey and the windows covered to cut reflections. They headed east along the tree-lined road that ran from Haze-brouck through villages, past dairy cows, hop fields and the ubiquitous red poppies, and on to Bailleul. Though there was little evidence of the war as they rattled eastward, the men were now only five kilometres from the front, and their 'excitement was intense'.[4]

Past Bailleul, half the company split off, some heading to St Eloi to work with the Royal Engineers Tunnelling Company, at that time struggling for control of the Mound, while another group headed north, to near Poperinghe, where they could learn from the Canadian 3rd Tunnelling Company working under-ground at the notoriously dangerous Hill 60.

Towards the front, they saw their first shattered building as they lumbered through Nieppe. Then the artillery positions became visible, the British gunners busily stacking shells ready for their next call to action, and in the distance, Armentières. Short of this large town, the grunting, smoky bus pulled to a stop. The men put on their packs, clicked together the brass buckles of their web belts and shouldered their rifles. Spreading out so as not to make themselves an attractive target to the ever-vigilant German gunners, the tunnellers marched into the main square of Armen-tières. Here, to their amazement, among the damaged buildings and rubble, was a crowd of inquisitive children who were uncon-cerned about the battle raging nearby and eyed off the dusty Australians 'with an interest generally shown to a Circus'.[5] It went against all the officers' preconceived ideas of the frontline. 'Do you wonder we asked ourselves the question, when are we going to get to this blooming War?' Woodward wrote.[6]

That night, Woodward and the officers enjoyed dinner with the Canadian tunnellers and soon 'found all conditions topsy-turvy to those we had imagined'.[7] They were surprised by the number of civilians who had remained in the city and the many businesses that were open and carrying on as if the war was far away. They strolled around town like tourists enjoying the warm night air.

Walking through the town square, known to the troops as Half Past Eleven Square because the shell-damaged town clock had stopped at that time, Oliver Woodward was led down a side street to the starting point of the communication trench that ran out to Houplines and on to the frontline. Here, in the still evening air, he could hear the sharp crack of rifle fire and the occasional rattle of a machine gun. An old Flemish couple passed by, arm in arm, quite unperturbed, out for a walk before returning to the relative safety of their cellar. On the way back, he stopped at a dimly lit café – its windows were boarded up, but the strong, sweet smell of brewed coffee and freshly baked cakes led him to it – then it was early to bed in the officers' billets at the rear of the company's headquarters. 'As I lay on my comfortable bunk I attempted to sort out reality from unreality but found it almost an impossible task,' he wrote. 'Tuesday 15th May 1916 can I rightly think be classed as one of the most confusing days ever experienced by me.'[8]

At 8.30 the next morning, he set out with the commanding officer of the Canadian tunnellers, Captain MacMillen, for an inspection of the underground workings at the section of the frontline the Australians were to take over. They drove to Motor Car Corner, a point about two kilometres behind the line beyond which motorised transport was not permitted in daylight hours. From there, the two officers set out on foot along the road that led to the head of the communications trench named Nicholson Avenue.

Suddenly, shells crashed to right and left of the road, throwing up fountains of dirt and sending whirring shards of metal through the air. Oliver Woodward had received his baptism of fire. He wrote:

Outwardly Captain MacMillen gave the impression of absolute indifference, but I have a suspicion that this mask was assumed solely for my benefit and actually he had the 'wind up' as much as I. The whole performance was so unnatural. Imagine the scene, two officers walking along a road on either side of which huge areas of earth suddenly leapt skyward, preceded by a screech and followed by a roar, and both officers appearing to ignore the whole performance.[9]

Reporting to the headquarters for the tunnellers – a dugout in a trench at a point known as Amen Corner – they were met by a young lieutenant, the officer in charge of this section of the line. Woodward looked about him. The dugout was only a little over two metres square and two metres high. Around the walls were timber planks, and sections of railway line held up a timber roof. Above this were sandbags and above those a layer of bricks covered with mud and debris for camouflage. Though it might save the occupants from shrapnel balls and shell splinters, it offered little protection from a direct hit or being buried alive.

From this tiny, muddy position, the daily coordination of the tunnelling work and the men was conducted. The tunnellers were divided into two teams and worked four days at the front then were rested for four days at billets in the rear. During their four days in the line, they worked six-hour shifts on a continual 24-hour basis: six hours on and then six hours off, in which they would eat and get what sleep they could. The officer and his sergeant would be on duty continually for these four days, grabbing what sleep they could and eating when food was available, and then only after the men had eaten.

Defending this part of the line were troops of the South African Scottish Regiment. 'I imagined that on the previous day I had got my full issue of surprises but here was a super surprise,' noted Woodward. 'Who would have imagined he would ever meet Dutchmen or Boers clad in kilts defending a Frontline Trench in France?'[10] After initially finding it comical, it dawned on Woodward that it was a sign of the strength and resilience of the Empire that men from so far afield and with such a recent antagonism to Britain were now working, fighting and dying for King and Empire.

Passing along the frontline, Woodward noted the high water table, which did not bode well for mining. Between the duck-boards laid in the bottom of the trench he could see dark, muddy water, and he could smell the musty, damp odour of wet hessian and timber. Water oozed between and out of the sandbags lining the sides of the trench and the parapet above, dripping endlessly into the quagmire below.

After a pot of tea at battalion headquarters, he was led down into some of the Canadian workings, known as the Monmouth

House and Essex Farm Mining Systems. Then, after a dinner of cold stew and tea, he set out again underground, crawling through tunnels just a metre high and less than a metre wide. Dank and wet, he dragged himself through a hundred metres of this tight tunnel system. Above, the nightly German strafe on the lines with trench mortars added to the unreality of the situation.

By the time he emerged it was dark. Back at the officers' dugout, he was offered a hessian bunk, one of three fixed to the wall, with the lowest just clear of the muddy floor. A hessian curtain, an attempt perhaps at privacy, flapped restlessly in the draught while water dripped from the roof and oozed between the timber that had been laid on the mud floor. But at least he had a bunk, he thought, unlike the men who were sleeping wet and cold on the fire step of the trench, protected only by a flimsy waterproof sheet. Wet, and physically and mentally exhausted, he stretched out fully dressed, placed his gas mask and revolver near to hand, pulled a mud-caked blanket over his head and was soon asleep.

Two days later, after a rest in Armentières, Oliver Woodward found himself back at Amen Corner, but this time with 40 men under his command. No sooner had he arrived than the battalion intelligence officer approached him and asked if he could investigate what appeared to be the sounds of enemy mining. Though this proved to be a false report, he felt for the first time the fear of a silently digging enemy tunnelling close to his lines, scraping away the earth beneath him or quietly packing bags of explosives. Any reports of active mining were coordinated by the intelligence officers and as well as a report going to headquarters such information was passed on to the troops and published in what was known as the *Corps News*.

At 3.30 am on 21 May 1916, as his men were working in the Essex Farm Mining System, a tunneller's pick struck the face, and a fall of earth revealed a black hole. They had broken through into a German tunnel. The men sent word, and Woodward raced to the shaft and pushed his way down the tunnel to the face. He and another officer, Lieutenant Allen, carefully broke down a further section of the wall, the wet mud falling at their feet, and gingerly peered in. They half-expected an explosion, a rifle shot or a smack in the face with a shovel, but all was silent.

They enlarged the hole, feeling a draught of fresh air blow on their faces. There was still no sound and no detonation, so they stepped through the hole into the enemy's workings for the first time. Around them, the tunnel was in a good state of repair and superior to the Allies': the walls were boarded and secured, and there was a good drainage system. Cautiously, with pistols drawn and each holding a small torch, their yellowish beams casting strange, distorted shadows on the timbered lining boards, they moved along the tunnel. Stooped, alert and frightened, they walked towards a tangle of boards and, bracing, found that the gallery had been badly shattered, probably by a Canadian camouflet. Access had been blocked. Why the Germans had not cleared this debris and reopened the tunnel was something of a mystery. For safety, Woodward established a listening post and returned to the surface.

Though they had been there just over a week, that same afternoon the Canadians officially handed over the tunnelling operations to the Australians and marched out. Two days later, they were relieved and returned to their billets in Armentières; they had survived their first spell at the front.

✕

The 2nd Australian Tunnelling Company, just south of Armentières, had taken over the operations of the 172nd Tunnelling

Company Royal Engineers. Though they had been in the area for only one week, they decided to slow down the Germans, who were very active underground there, by firing a camouflet on 22 May 1916. It was the first to be fired by the Australians on the Western Front. They followed this up with another camouflet when they heard Germans very close to their gallery in the early hours of 30 May. Soon after, smoke was seen rising from the German trenches, most likely through the entrance shaft to the mine, indicating a successful attack.

In late May 1916, in an assault on the Australian line near Cordonnerie, just to the west of Lille and near Fromelles, a German raiding party, including four miners who carried small explosive packs, intended to destroy Allied dugouts and the entrances to mining operations. Members of the 2nd Australian Tunnelling Company were caught up in the fighting when the German raiders mistook the entrance to an Australian mine for a large dugout and threw in an explosive charge. The Australian tunnellers and pioneers rushed out. Private William Edward Cox, a labourer from Parramatta, was the first to emerge. He was shot in the stomach at close range and fell back, dying. The next man out was bayoneted but survived; he and four others were captured.[11]

Meanwhile, the 3rd Australian Tunnelling Company was near Lens. In June they fired a pipe mine, a common method of making quick, localised attacks on the enemy's line. In essence it was an explosive charge fired at the end of a bore hole that had been driven towards the enemy lines, but it generally had limited and unimpressive results.

The Australians were very welcome at the front. They brought wide and varied experiences in tunnelling and new ideas, and they had arrived at a time when it was extremely difficult to get more miners from the British pits, especially from the coalmines, as men were pressed to keep up coal production for the war effort.

By now, the firing of camouflets and large offensive mines had become incessant, reaching its zenith along the British 1st Army front from Vimy Ridge, near Arras, north to Laventie, which is northwest of Fromelles, a distance of about 30 kilometres. During June 1915, the British fired 79 mines there, while the Germans fired 73. Along the whole British front, 227 mines were fired that month, about one every three hours.[12]

×

In late May, the New Zealand Tunnelling Company arrived, and they were attached to Woodward's company for training. It was a case, according to Woodward, 'of the blind leading the blind'.[13] The unit had been recruited in September 1915 from the mining areas of New Zealand, and they trained in their home country and Falmouth, in England. After getting frontline experience with Woodward's men, they were sent to Vimy Ridge to take over underground operations from the French.

There were no targets on the German lines opposite Woodward's section of the front that were worth an offensive-mining effort. And the area was not suited to offensive mining, as it had a high water table – six metres – beneath which the men could not dig. So the tunnellers focused on defensive mining operations to protect the trenches. This meant digging a vertical shaft or an incline to a depth between three and six metres below the surface and then a level gallery or tunnel towards the enemy. Below no-man's-land, a lateral gallery was dug parallel to the Australian frontline, and from this, galleries spaced 15 metres apart were driven towards the German frontline to form listening posts. From the end of these galleries two short tunnels were driven left and right, forming a 'T' to ensure no aggressive German mining could be carried out and not detected. And to be prepared for German tunnellers, bore holes of 20 centimetres diameter were drilled and filled with explosives.

While tunnelling progressed, Oliver Woodward, some officers and a few select men attended the newly established Mines Rescue School held at the Headquarters of the 171st Royal Engineers Tunnelling Company, outside of Nieppe, a few kilometres to the northwest. There they would have been trained in the use of Proto kits, mine rescue and listening equipment, boring machinery and explosives, the removal of enemy mines and camouflets, and mining tactics.

Such training was invaluable because the Australian tunnelling companies would be called upon to assist with trench raids and other offensive actions when their specialist sapper or explosive skills were required. And, soon after his return from attending the mines rescue course, Oliver Woodward was nominated by his commanding officer for just such a job.

Twelve

The Red House

To the northwest of Armentières, past Houplines, is the small village of Touquet. This village was part of the operational area for the No. 1 and No. 2 sections of the 1st Australian Tunnelling Company, which at the time had its headquarters in Armentières. Here, just outside the German wire entanglements, about 80 metres from the Allies' trench, was a partly damaged farmhouse, known as the Red House, from where German machine gunners and snipers were firing daily, causing Allied casualties. It was Woodward's task to cross no-man's-land, demolish the Red House with explosives and eliminate the German gunners. This was an especially dangerous operation, and Woodward later wrote:

I frankly admit I experienced a sinking sensation when this news was given me. Up to this date I felt quite a hero to be even in a Frontline Trench. Here was a job which would test my nerves in full. It was a critical stage in my career. A failure and I would be damned, a success and I would win my spurs as a soldier. Which was it to be?[1]

He returned to his unit and selected two men to undertake the task with him: Sergeant Hugh Fraser and Sapper Morris. The men's enthusiasm when they were told of their selection fortified and steeled him. 'I selected as my companions for the patrol, Sergeant Fraser and Sapper Morris, and the joy with which these men received the news helped materially to quell my fear of the unknown,' he wrote.[2]

They packed a charge of 46 kilograms of high explosives into a canvas bag and attached two sets of fuses so that the explosives could be detonated by either an electrical charge or by a length of instantaneous fuse, which was lit with a match. Woodward allowed nearly double the distance between the trench and the Red House for the length of the electrical lead, to account for the roughness of the ground, the shell holes and the fact that he knew he would not travel in a straight line in the dark. The plan was to crawl out, lay the charge beside the ruined walls, crawl back and into the safety of their own trench and ignite the charge with a small Military Portable Exploder, a hand-held device that once pushed down sent a current down the electrical lead to explode the charge.

In the dusk light of 10 June 1916, Woodward and his two men left their headquarters. Four hours later, after picking their way along the winding communication trench to the front, they arrived at the headquarters of the West Kents, the English regiment holding this part of the line. Here he picked up four men who would provide flanking protection, two on each side of his party as they crossed no-man's-land. After being briefed, the rifles, explosive charge and electrical leads were checked and the men sat down to await zero hour, which was midnight, half an hour away.

Woodward found a small, unoccupied recess in the trench wall away from the party and sat down. Of his thoughts he later wrote:

We were to carry out a duty in the sight of a large number of the manhood of our nation. To show signs of cowardice in any form in the hope that thereby one's life might be preserved meant that life under such conditions would be purchased at a price too terrible to contemplate. There was only one course open and that was to overcome any sign of cowardice and this I strove to accomplish.
I can honestly say that throughout my life as a soldier I never really overcame Fear. It was always present to a greater or lesser degree.
I believe I did conquer cowardice. I am of the opinion that those who claim to never have experienced Fear in War were either braggarts or were devoid of intelligence. A hard statement but none the less true. The degree of heroism was the degree by which individuals conquered cowardice.[3]

Because nights were short at this time of the year, zero hour had been set for midnight. Hearing the occasional burst of machine-gun fire from the German trenches, the raiding party slipped over the parapet and crawled towards the barbed-wire entanglement that stretched along their front. They wriggled through a small gap in the wire, pulling themselves along by their elbows, their bellies dragging through the mud and their noses just clear of the stinking ooze.

Suddenly a flare rose skyward, leaving a trail of sparks and making a fizzing noise like a cracker-night skyrocket. The men froze. Woodward pressed his face into the earth, holding his breath and waiting for the machine gun, now less than 50 metres away, to open up. Surely the Germans must see him and his two men lying prone behind him.

The flare fizzled out and died, its burning parachute falling to the right somewhere. Just then, a machine gun started up further down the line, its bullets sweeping no-man's-land in a wide arc high over his head. He remembered the heavy bag of explosives he carried and the effect of the impact of a bullet.

Closer to the German lines the men came across grass, which gave them a little cover as they crawled forward. Their advance was slow as they wound between shell holes and regularly stopped to check their direction and progress. In the glow of a far-off flare, Woodward saw the silhouette of the Red House, now quite close and ominous. Gathering the men around him, he lay still for 15 minutes, straining to hear the faintest sound from the ruins that might betray a German presence. All was quiet.

Carefully he transferred the heavy charge from his back, cradling it against his chest like a baby. He could smell the mildewy canvas and the explosives, and feel the insulated wires. Leaving the men behind, he crawled forward into the shadow of a wall of the Red House, and again waited and listened. He could hear German voices, low and whispered, but they were coming from somewhere to his left and from the German frontline trench, behind the house.

Slowly he rose and reached for the top of the wall, planning to climb over and drop down onto the rubble-strewn first floor of the house. But as his fingers gripped the crumbling wall, cement render broke off in his hand and fell down through a gaping hole in the floor, into the cellar. Had he swung over the wall as planned, he would have found himself ten metres down in the cellar, injured and easily captured. He slowly retracted his hand and paused. The murmuring voices continued.

Hugging the brickwork, he strained to get the bulky bag of explosives up and over the wall, allowing it to hang and swing gently in mid-air high above the cellar. Maintaining his grip on the wires, he gradually lowered the bag to the cellar floor, where it came to rest with a quiet bump. Flares still soared high into the air

Oliver Woodward (left) at the Laloki copper mine near Port Moresby with the geologist Colin Fraser (later Sir Colin Fraser). *Courtesy Woodward family*

Oliver Woodward as a young man prior to the war. *Courtesy Woodward family*

Marjorie Waddell as a young woman, 1915. *Courtesy Woodward family*

Woodward as an officer in the AIF, 1916. *Courtesy Woodward family*

HMAT *Ulysses*, the troopship Woodward and his men travelled to France aboard. *Courtesy Australian War Memorial*

The high ground that was Hill 60 from the British lines.
Courtesy Ross Thomas

Tunnelling on the Western Front, 1916. *Courtesy Ross Thomas*

British sappers at work in a chalk tunnel, probably troop accommodation as denoted by the size of the tunnel. *Courtesy Ross Thomas*

John Norton-Griffiths on the front, 1915. *Courtesy Ross Thomas*

A German boring machine under the supervision of an officer. *Courtesy Ross Thomas*

A German mining listening gallery in early 1917.
Courtesy Ross Thomas

Two German soldiers with hand grenades at the ready in a trench on Hill 60. *Courtesy Australian War Memorial*

Miners of 1ATC excavating a headquarters dugout 25 feet below the surface near the Menin Road, Ypres Sector, September 1917. *Courtesy Australian War Memorial*

Members of the 1ATC excavating troop accommodation close to Hooge Crater, Ypres Sector, September 1917. *Courtesy Australian War Memorial*

The entrance to the Wallangarra Dugout, also known as the Catacombs, built by the 1ATC at Hyde Park Corner, Ploegsteert Wood. Note the kangaroo above the entrance. *Courtesy Australian War Memorial*

The 15-inch Howitzer that attracted enemy aerial bombing to the position of the 1ATC camped nearby, Menin Road, Ypres Sector, October 1917. *Courtesy Australian War Memorial*

Australian pioneers building a plank wagon track over recently captured ground, Chateau Wood (near the Menin Road) September 1917. *Courtesy Australian War Memorial*

The Lille Gate, Ypres. Australian tunnellers built accommodation and headquarters into these damaged ramparts. The destroyed town of Ypres can be seen in the background. *Courtesy Australian War Memorial*

Woodward being presented with a second bar to his Military Cross by H. R. H. The Prince of Wales in the grounds of Government House, Adelaide, 13 July 1920. *Courtesy Woodward family*

Officers of the 1ATC who were responsible for the firing of the mine at Hill 60. Here they are pictured at the Malhove rest camp six weeks after the big event. *Back row from left:* Lieutenants John Royle, James Bowry and Hubert Carroll. *Front row from left:* Captain Oliver Woodward, Major James Henry and Captain Robert Clinton.
Courtesy Australian War Memorial

Oliver and Marjorie's wedding, Brisbane,
3 September 1920. *Courtesy Woodward family*

Woodward at his desk at the Port Pirie Smelters in the late 1920s.
Courtesy Woodward family

1ATC Lieutenant John Royle, photographed in 1934
with one of the electrical exploders used to fire the
Hill 60 mines. *Courtesy Australian War Memorial*

Woodward, April 1943. *Courtesy Woodward family*

A small wooden box handmade by Private Tiffin from timber salvaged from the destroyed Ypres Cathedral and Cloth Hall. Woodward's Military Crosses are displayed. *Courtesy Woodward family*

and popped, blanketing the area with an eerie glow, and he was amazed that he or the canvas bag had not been spotted. The voices focused his mind: rather than return to the relative safety of the trench to detonate the charge, he must blow it now, close to the German line. 'In view of the fact that we could still hear voices it seemed to me that there was a possibility of the charge being discovered at any moment,' he wrote. 'I therefore decided to fire by means of the ignition fuse.'[4]

Woodward crawled back to the men and signalled them to draw into a tight circle. He lit a match to ignite the fuse, hiding the flame from the Germans with his helmet. Then, bent double, he and the men scurried a few short steps to take shelter in the darkness of a shell crater and waited. But nothing happened. There was no explosion. It could have been the rough handling as he dragged the canvas bag across no-man's-land, or perhaps the detonators had broken free when the bag had landed on the floor of the cellar. He had one last chance: the electrical fuse. He would run the electrical cable back to the Allies' trench and detonate the explosives from there.

He crawled back to the brick wall of the Red House and felt the electrical cable to make sure that it was still in place and secure. With his two men ahead of him, he worked his way back to the trench, carefully playing the cable through his fingers until he felt the wire break away and fall to the ground. He twisted the wires roughly together and hoped the connection would hold. Moving on again, he found five more breaks in the wire. The wire ended 20 metres short of the safety of the trench.

Undeterred, Woodward sent Morris back for the plunger. Then, with his sergeant and the English troops safely to ground, he joined it to the wire and drove the plunger handle down. Immediately there was a terrific explosion, and bricks and debris flew high into the air and crashed about them. Grabbing the exploder, he raced at top speed for the gap in the wire, wriggled in and threw himself headlong into the trench. The explosion was the signal for the British artillery and the British troops to open fire on the German lines, and Woodward and the men made it back just in time as the front came alive.

For his 'conspicuous gallantry' that night, Woodward received the first of three Military Crosses he would win during the war – an honour that made Woodward uncomfortable because it set him apart from his men:

> The trophies of the chase to the officer, nothing to his men, other than the satisfaction of knowing that they had done their duty equally as well as the officer. I was quite embarrassed when my companions on the stunt offered me their hearty congratulations and I begged them to realise that I was the victim of over enthusiasm and lack of discretion on the part of Senior Officers.[5]

Sadly, we don't know what became of Sapper Morris, but we do know that Sergeant Hugh Fraser also went on to be decorated, winning a DCM for rescuing buried men under heavy shellfire. He died on 31 May 1918 at the Casualty Clearing Station at Ebblinghem, on the main road between St Omer and Haze-brouck. A wiring party he was leading came under shellfire, and he received a wound to the abdomen. He died early the next morning and is buried in the Ebblinghem Military Cemetery.

THIRTEEN

The First Big One

By the time the Australian tunnelling companies had arrived in France, preparations were all but complete for the great offensive planned for 1 July 1916 on the Somme. When planning had begun in December 1915, it was to be a predominantly French offensive; they were to commit 40 divisions as opposed to the British commitment of only 25. But when French reinforcements had to be diverted to Verdun to deal with the German attack, the British became the dominant partner. The French were now offering only five divisions.

The Somme valley and the watershed formed by the basins of the Scheldt and the Scarpe rivers had confronted the British planners with a very different landscape compared with Belgian Flanders. Here the land rose to a height of 150 metres and the chalk offered a safe haven for the defending Germans, who held the heights in an arc to the northeast of Albert.

The frontline designated for the attack stretched 22.5 kilometres from Gommecourt in the north to Maricourt in the south, all well defended by major German fortifications. They could not go through them and they could not go around them so they were left to try to go under them. Their strategy for this underground war had two aspects: the first was to mount a massive mining operation in which mines bigger than any ever used before would be blown underneath the German line. The concern was the inability of British infantry to take advantage of the mines as they had a poor track record, especially after the disasters at St Eloi and the Hohenzollern Redoubt. The Germans were the masters of counterattack, and British tactics in crater warfare never matched theirs. It was one thing to successfully tunnel under the Germans, blow a mine and destroy their front-line trenches, but unless British infantry could take the newly won ground and hold it against German counterattack, the resulting crater simply gave the Germans new defensive positions and a more commanding high point over the British lines.

That was where the second element of the mining operation came in: the technique of the 'Russian sap', something developed in the Crimean War and used successfully against the Turks at Gallipoli. This meant digging shallow tunnels, about two metres in height with only half a metre of earth above, towards the German frontline. The British troops could advance through them undercover towards the German trenches and, at an oppor-tune moment, break out and charge over a shorter distance towards the German frontline. These saps could then be enlarged and strengthened to become a communication trench to the newly won positions, as well as providing 'emplacements' for both Stokes-mortar crews and machine guns. However, unlike Gallipoli, the opposing frontlines were not just a hundred metres or less apart, but in some places nearly a kilometre apart. This required serious planning and dangerous, exposed digging.

The British planners also had to solve the problem of the debris scattered over a wide area by massive mines. If the attacking troops were too close, they suffered casualties from the falling debris, but if they were too far back, safe from the debris storm, they gave the defenders time to gather their wits and man their machine guns. The tunnelling officers came up with a new idea: they would increase the explosive charge in the mine (it was called 'over charge'), which would throw more debris higher into the air, extend the radius of the falling material (killing more of the enemy in the process) and create a higher lip to the crater, providing better protection for the attacking troops. Usually the lip was about two metres high, but if they could create a three- or four-metre lip, it would provide significant protection for their attacking troops.

To cover the 22 kilometres of front, the British needed the help of the tunnelling companies, so five were sent to work in this sector. And to do the work needed by 'zero hour', the tunnelling companies pulled in every available man they could, in some cases doubling their usual number. The British 252nd Tunnelling Company at the very northern end of the line had more than 2000 men; many of them attached infantry who probably hated the idea of being 'tunnellers'.

Opposite the 'Redan', the northernmost mine, the Russian saps started to be pushed forward from April 1916. The Germans constantly watched no-man's-land and located one of the nine saps on the front north of Beaumont Hamel. They fired a small charge, killing seven British tunnellers. A week later, they fired a second charge killing another two tunnellers, but work went on and the saps were ready in time for the offensive.[1]

In this same area, the tunnellers had placed two mines; one at a depth of 17 metres and the other 19 metres. Just south, a gallery had been driven at Beaumont Hamel into the hard chalk, but the tunnellers had struck large chunks of flint and this had slowed them considerably. Men had to slowly and silently prise the flint from the chalk with bayonets after the face was thoroughly wet with water, then catch the falling stones so as not to make noise and alert the Germans.[2] Once they had driven 320 metres out and under the German defences, a chamber was dug and packed with 18.5 tonnes of ammonal.

And all along the line, with only a flickering candle for comfort, thousands of men sweated at the chalk walls before them, clawing at the face with bayonets and bare hands, wondering when the next German camouflet would explode. They all knew the results of a blast. They had seen it: liquid, oozing chalk that ran red, pieces of human flesh clinging to the tunnel walls, perhaps a hand protruding from the white, gluey bottom. The men knew death was everywhere, imminent and close, and that there was no way, in this deep, black and uncharted hell, their mothers would visit their grave. Dante's inferno was very real and they lived it.

Further south, opposite Thiepval, now the site of the massive British Memorial to the Missing of the Somme, ten shallow saps were pushed out and emplacements constructed for the positioning of heavy trench mortars. One sap, 'Inverary', went out 75 metres to within 25 metres of the German line, and another, 'Sanda', was more than 100 metres long.[3] It actually connected the German trenches and later became invaluable for communications. Saps similar to these were pushed forward all along the British frontline and one sap driven opposite Ovillers, named 'Rivington', reached just ten metres from the German line. It was so close that from the end Germans could be heard talking and laughing. The other sap, 'Waltney', dug into the chalk, was pushed out over 500 metres.

The area around the small village of La Boisselle was very active with mining. Many shafts were more than 30 metres deep and had galleries at four different levels. The big mine south of the road was 'Lochnagar', which had a steep incline sunk to nearly 30 metres and two chambers packed with 16.3 tonnes and 11 tonnes of explosives. These tunnels had been driven in December 1915, along with the sap 'Kerriemuir', which connected the Lochnagar crater to the German and the British frontlines. After 1 July, this was used extensively by troops and a whole battalion passed through it. Later it was used for the evacuation of the wounded.[4]

Opposite Fricourt, a number of deep shafts were dug, but this area came under heavy and constant artillery attack and many of the entrances were closed. Here too the chalk was very hard and the only way to tunnel through it was to drill six-centimetre holes. Four holes were drilled in a circle known as a 'round' then filled with an explosive, generally a small charge of 'Blastine', and ignited. The Germans adopted a similar method and their rounds could be heard exploding far out in the chalk. Three Russian saps were driven out, this time to be used as flamethrower positions, but only one was subsequently used as the German line was not captured in this area.

Towards the Somme, at the southern end of the line between Fricourt and Mametz, the enemy was close and the firing of large camouflets was a regular occurrence. Everywhere the tunnellers were like busy moles, pushing out galleries and Russian saps, building underground accommodation for the troops, loading deep chambers with tonnes of ammonal and preparing push pipes.

One sap near Casino Point came so dangerously close to the enemy it was decided to silently drill forward with an earth auger. Two tunnellers, lying in the cramped space at the end of the sap, silently worked their auger and bagged up the spoil. Progress was slow and delicate, as they calculated they were only ten metres short of the German line. Suddenly, however, at eight metres, they broke through into a German officer's dugout. Withdrawing their auger as quickly and quietly as they could, they lay panting, fearing swift retribution. But nothing happened. As they lay there, they could hear German voices from time to time and footfalls on the duckboarded floor, yet they appeared not to have been noticed.

The tunnellers stopped work on their sap and crawled back to report the incident to their officer. Fearing the discovery of the sap, he crawled out along the shallow tunnel to investigate and found that the Germans had seen the hole and simply plugged it up. Lying close to the auger hole he listened, but he couldn't hear anything suspicious and there was no indication that the Germans realised what the hole meant. The next day the tunnellers heard picks at work. Again, nothing came of this and the work soon ceased, so the tunnellers continued undisturbed and prepared a charge in readiness for the attack. This they then fired later, destroying the enemy trench and two dugouts.[5]

Everything was in place for the offensive on the Somme, but some in Allied High Command had reservations about it. Haig was reluctant as from the earliest days he had wanted to attack on the Ypres salient, regain the high ridgeline from Messines to east of Ypres, and occupy the ports of Ostend and Zeebrugge now held by the Germans. Lord Kitchener was opposed to the large-scale Somme attack, preferring to continue making small offensives. Even France's General Foch was questioning whether the offensive should take place. For the first time the word 'attrition' was coming to replace phrases like 'total victory' and 'breakthrough'.

For seven days leading up to 1 July 1916, the British subjected the Germans to a relentless bombardment of about 1.7 million shells. The Germans were expecting an attack and their machine guns were oiled and ammunition belts ready – the very equipment that would turn this day into the most disastrous in British history.

One enormous blunder was the timing of the detonation of the first mine, at Hawthorn Redoubt, north of the River Ancre. Initially it was planned to fire this mine four hours before the attack, but General Harvey argued that it would simply alert the Germans. The compromise was to fire it ten minutes before zero hour. The explosion proved devastating, utterly destroying the redoubt and along with it many Germans, but it did exactly as Harvey had feared. All along the line, the enemy came up from their underground shelters and lined the parapets, mounted their machine guns and laid out their bombs. The Germans were ready and waiting.

Then, at 7.28 am, two minutes before zero hour, eight large mines were blown, creating massive cavities and throwing up high rims of debris.

Near Lochnagar, well over 100 metres of German frontline trench was obliterated, leaving a crater 100 metres across and 30 metres deep. At Tambour, near Fricourt, one mine failed due to moisture, but two other mines devastated the German line. At nearby Casino Point, the mining officer watched the seconds tick down on his watch, ready to push the electric plunger home. To his horror, with minutes still to go, he noticed British troops climbing over their parapet and heading towards the German line. There had been a problem with the synchronisation of watches and the men had started out too early. What was he to do? Quickly he forced the plunger down, and the German frontline vanished, but not without inflicting casualties on the advancing troops.

At zero hour, the whistles blew and the men filed out of their trenches, crossed their start lines and headed out into no-man's-land, their rifles across their chests and advancing in lines 100 metres apart. The preceding seven days' bombardment had supposedly cut the wire and wiped out the German troops and frontline defences, but the commanders had not taken into account the deep shelters the Germans had dug underground, and the discipline that meant they could quickly take up positions along a tattered line and return fire.

In many places the commanders hadn't told their men about the Russian saps. This lack of communication, put down to the great secrecy surrounding the operation, sadly translated into the death of probably hundreds of British soldiers. There were times when tunnelling officers yelled to advancing troops just a hundred metres away to use their saps, but they could not be heard in the noise and confusion of battle.[6]

Despite all the hard, perilous work of the tunnellers, the British regiments walked, line upon line, into the enemy guns. The lines simply crumpled, leaving no-man's-land 'strewn with khaki figures mown down in swathes like ripe corn before a scythe'.[7] At the end of the first day, nearly 20,000 British were dead and 40,000 wounded.

The saps were put to good use in a mere handful of places, for the resupply of forward troops, the passage of runners and re-inforcements, and the return of the wounded. One battalion commander relied on saps to provide protection for his men at an exposed and dangerous salient, allowing them to hold the newly won ground.

But in the northern sector of the line, the use of a Russian sap 25 metres from the German frontline proved nightmarish. The men had expended their supply of mortars within eight minutes, and being so close to the Germans, they started taking casualties from grenades thrown over the short distance.

The southernmost mines, at Fricourt, Mametz and Montauban, had caused great damage while shattering the defenders. Here the tunnellers were able to inspect the German mine systems. One bright spot was the discovery of the very German practice of marking the daily progress of the mine construction on the timber walls, which confirmed for them that the Germans' daily progress underground was perhaps two-thirds that of the British tunnellers'. This may have been due to the high standard of workmanship, which the Allied tunnellers agreed was probably 'more elaborate than essential'.[8]

In the tunnels at Fricourt, they found the diary of a German mining officer. It showed how ineffective and harmless the British countermining had been and detailed the procedure the Germans had developed to successfully clear their men from their workings before an Allied mine was blown. Here too was evidence in the form of transcriptions, notes and numerous intercepted messages spread about the tunnels, along with the until now secret equipment they had used to record British telephone messages, all seriously disappointing news to the exhausted men.

But what probably affected the men the most was the discovery of very deep shafts, something they had not expected. A German gallery was found to be more than 60 metres deep. Although this was not an effective depth for offensive mining operations since they would simply need too much of any explosive known at the time to reach the surface, it did cause concern for the Allies, as they thought they had deep mining operations all to themselves.

What the British probably did not appreciate at the time was that German deep mining experience was very limited, as were the number and breadth of experience of men in the German tunnelling companies. Many of these men were detailed to these companies against their will, and in some cases they were directed by German officers from the surface rather than underground, possibly because they lacked mine training.

Even though the Battle of the Somme was an abject failure, it was the turning point in the tunnelling war. The mines themselves had all gone off successfully, proving the viability of deep mining. The British were now taking the mining initiative and were trying new methods, new equipment and new tactics. And the Australian and other colonial tunnelling companies were contributing different skills, experience and ideas. From now on, the war underground would be a very different fight, and one in which the Germans would never again achieve the ascendancy.

By the middle of July, Allied casualties had reached 100,000. After partial strategic success, particularly in the areas along the French front, the battle quickly bogged down and the spreading new trench lines again sucked in more men and material. Three Australian divisions (the 1st, 2nd and 4th) were ordered south, to the Somme. The Germans, too, were moving men there, so the British decided on a diversionary attack further north, opposite the German-held village of Fromelles. The British 61st Division and the untried Australian 5th Division would launch the attack. Some of the 3rd Australian Tunnelling Company was moved to an area behind the British lines near Fromelles. Here they used push pipes and ammonal charges to create shallow trenches to provide cover for the attacking troops as they charged across the flat expanse of no-man's-land. These, like the whole of the Fromelles attack, were a failure. The 5th Division was virtually wiped out in this brief action, taking 5533 casualties, of which 1780 were killed.

FOURTEEN

Sojourn at Ploegsteert Wood

The Australian tunnelling companies were still to the north of the Somme battlefields, spread along the line around Armentières, when on 30 June 1916 the Germans laid on a heavy bombardment of the town. It was said that a German-born New Zealand soldier had deserted to the enemy and provided information about the Allied divisional headquarters and men's billets. These were targeted, resulting in heavy casualties, and Oliver Woodward was lucky to escape unharmed.

The shelling hastened the planned transfer of Woodward's unit to a new camp that had just been completed in the Ploegsteert Wood sector, to the north of Armentières, just over the Belgium–France border. The wood, about two-and-a-half kilometres long by a kilometre wide, was below the southern end of the Ypres salient, just four kilometres south of Messines. The 2nd Australian Tunnelling Company was at Fleurbaix, south of Armentières and just to the west of Fromelles, while the 3rd was at Red Lamp and the villages along the front at Aubers, very close to Fromelles.

Over the years, Ploegsteert Wood had been heavily shelled.

Not long after he arrived, among the shattered trees on the eastern side of the wood, Oliver Woodward observed a photographer and artist closely studying one particular tree, making sketches and taking photographs – an odd thing to do at the time, he felt. His curiosity was satisfied some days later when he found that the tree had been sawn down and a replica tree, made of steel and very accurately painted like the original, had replaced it. At the bottom was a small trapdoor through which an artillery observer could squeeze, and then, upon climbing the interior ladder, find a safe high point from which to observe the German lines and direct fire.

Three more Australian tunnelling companies, the 4th, 5th and 6th, had been formed. When they arrived at the front they were amalgamated into the three existing companies, bringing their numbers to more than 500 men each.

Half of the 1st Australian Tunnelling Company was undertaking offensive mining at Trench 123 and the Birdcage, just east of the wood. Oliver Woodward was disappointed that instead he and his men had been given a defensive mining job: building a large system of dugouts between the frontline and the support line at Prowse Point, just to the north of Ploegsteert Wood. The British High Command had finally woken up to the importance of deep shelters for headquarters and the protection of men. Unlike the Germans, who realised that men survived shelling and the harsh weather when underground, the British had the attitude that underground shelters made men lazy and soft, and that when the requirement came for them to fight, they would not come out and engage the enemy. Unbelievably, the British High Command had been happy to see their men standing around in the snow and in mud-filled trenches, suffering trench foot, frostbite and the impact of shelling – just as long as they did not get soft.

But now they were beginning to see the benefits of underground protection, and with greater numbers of tunnellers on hand, the British stepped up the digging of deep dugouts from rear areas to the frontline. It was to this effort that Oliver Woodward and the tunnellers were put to work at Prowse Point.

Woodward had just settled into the routine at Prowse Point and his men were well under way with the construction of a massive dugout, when on 8 July 1916 he was ordered to report to the 10th Queen's Regiment in the Le Bizet sector, to the south and just north of Armèntieres.

He had achieved a certain degree of fame for demolishing the Red House, and nearby and along the same section of the line, the Allies needed to take out a similar structure known as Machine Gun House. This was a ruined farmhouse about 20 metres in front of the German frontline, in which the Germans had set up a fortified machine-gun position. It was only about 250 metres to the north of the Red House and situated on the railway embankment near Rabeque Farm.

That night, a raiding party of 30 men, under the command of Captain Pillman, was to attack the German trenches opposite Machine Gun House. At the same time, Woodward was to command a party who would approach the house and demolish it with explosives.

At 11.30 pm the raiding party moved out, slipping through their own wire and advancing slowly across no-man's-land. Close by the German wire they were discovered, and machine-gun and rifle fire ripped into the raiding party. Almost immediately, Captain Pillman dropped wounded into a shell hole. The men, given the order to retire, raced back to the safety of their own trenches.

Soon after, a badly wounded sergeant dragged himself in and

reported that the captain was wounded and lying in the shell hole where he had fallen. Woodward was with a New Zealand officer when he heard the news. 'Instantly we both saw where our path of duty lay and conquering cowardice but not Fear, we left the trenches to give assistance to the wounded comrade,' he wrote.[1]

The failed attack had woken half the German army. Machine-gun fire played across no-man's-land, and the sky was lit with flares and the flash of explosions. Fearful of their chances, the two men leapt from the trench and raced into the blizzard of bullets, calling for Captain Pillman as they went. They found him with a deep wound in his thigh. He was unable to walk, so they lifted him onto Woodward's back and set off across the rough, broken ground. 'I shall never forget that trip across No Man's Land. That we were not shot to pieces seemed a miracle. Bullets cracked in our ears as if they had missed by a fraction of an inch. We could make no effort to dodge being seen when the flares went up,' Woodward wrote.[2]

They somehow made it back to the safety of their own trench and placed the badly wounded man on a stretcher. But his injury was even worse than it had first appeared.

Here Captain Pillman gripped our hands and thanked us for what we had done and a few minutes later he died. The bullet had been deflected into his body. I treasure above all else of my war life his expression of joy and relief when he found we had come to his assistance and his handgrip and word of thanks given to us just before he lost consciousness.[3]

Woodward returned to Ploegsteert Wood and worked with his men on the Prowse Point dugout. Here they could work without fear that the Germans would hear them or blow camouflets. But they were still not far from danger, as Woodward found when he had his first experience of gas in mid-July 1916. 'Fortunately this one was of our own creation,' he wrote.[4] The Allies had bombarded the Germans with gas shells and 'so annoyed the enemy that he gave strong retaliation'.[5] Ploegsteert Wood became infamous for gas attacks, and many a soldier's diary mentions the black, grotesque bodies of men that lined the quaintly named 'streets' – Hampshire Lane, The Strand, Regent Street and Oxford Circus – that ran through the wood.

Woodward's section was under pressure to complete the dugouts quickly, and when several of his fellow officers became sick he worked for three weeks straight.[6] It was not until 2 August that he was able to return to the rest camp and take a bath. He said in his diary, 'By this time I had almost qualified for admission to the Society of the Great Unwashed.'[7] At the rest camp he would have had his clothes deloused and washed and had time to make repairs and write letters home. While here, he received a letter from his father: the army had told him that Oliver, after enlisting in October 1915, simply vanished. They had no record of his training or posting, and could not trace him. For Woodward, this was quite a shock.

I received a letter from my Father advising that he had been informed by the Military authorities that his son Oliver Holmes Woodward had enlisted in October 1915 and had not since been traced. I felt hurt that a graduate of the Moore Park Officers' Training School had been lost sight of so easily. As this course consisted largely of Naval work such as knots and lashings I presume the authorities searched for me in the ranks of the Navy before appealing to my Father.[8]

Despite Woodward's humorous dig about the authorities searching for him in the navy, he must have found it galling to be risking his life at the frontline while some in the army were unaware of his presence and had implied to his family that he had deserted. Presumably this was swiftly sorted out, however, as Woodward makes no further comment about it in his diaries.

It was around this time that some unfortunate news arrived at the headquarters of the Australian tunnelling companies. Their now famous and highly respected geologist, Major Edgeworth David, had had a serious fall on 6 October and been sent to London.[9] David, they learnt, had been inspecting a well near Vimy and, while sitting on a wooden board and being lowered down by windlass, the rope broke and he fell 25 metres down a mine shaft. He crashed to the bottom but recovered consciousness quickly, and was able to call for help. A doctor arrived, bandaged up his bleeding head and secured him to a new line. Not one to be defeated, as he was being hauled up he called to the rescuers above: 'Pull me up slowly and give me time to observe the water level in the shaft.'[10]

Edgeworth David had been severely injured. He had a deep cut to the head, broken ribs and internal injuries. Though he was never to fully recover, after only six weeks in hospital he returned to his duties, but was forbidden from going within one kilometre of the frontline without his boss, General Harvey's, permission. General Harvey had been to see Edgeworth David in hospital and had been horrified by his accident and injuries. He understood the brilliance of the man and the need for more geology expertise on the front, so he decided that rather than compromise Edgeworth David's important contribution to the Allied tunnelling effort, he'd simply restrict his movements from then on. As a result Edgeworth David was usually to be found, much to his annoyance, with Headquarters staff safe behind the line.

In June 1917, Edgeworth David was promoted by General Harvey to Chief Geologist, attached to the Inspector of Mines at the General Headquarters of the British Expeditionary Force. He was awarded the Distinguished Service Order, was mentioned in dispatches on three occasions and promoted to the rank of lieutenant colonel. While he was at the front, his son won a Military Cross while serving as a Regimental Medical Officer with the 6th Cameron Highlanders and his daughter served as a driver with the Women's Army Auxiliary Corps.

After the war, he was made a Knight of the British Empire and became known as Sir Edgeworth. He resumed his work at Sydney University, but his main focus was to complete a comprehensive geological map of Australia. To do this, he took leave from the university and travelled around the country collecting samples and mapping the geology of the continent. It was published in 1932 as *Geological Maps of the Commonwealth of Australia*.

Edgeworth David, a popular and celebrated academic, was hero-worshipped by his students and attracted many of the best minds to geology. During his life he was awarded numerous honours and prizes including honorary doctorates from the universities of Wales, Cambridge, Sydney and Manchester and the highest recognition from scientific and geological societies worldwide. He died of pneumonia in 1934 in Sydney and was given a state funeral at St Andrew's Cathedral before being cremated with full military honours.

✕

With the completion of the Prowse Point dugout on 18 August 1916, Woodward's No. 2 section was moved to nearby Hill 63, close to Hyde Park Corner, to start work on a new system of dugouts. Woodward was ordered to take an intensive ten-day course at the 2nd Army Mine Rescue and Mine Listening School, at the Headquarters of the 177th Royal Engineers Tunnelling Company, at Proven, about 30 kilometres to the northwest and well behind the line. A model mine had been constructed at the school, complete with shafts, galleries and tunnels. This model allowed for the training of tunnellers in different types of soils, and gave them the opportunity to hear manufactured sounds underground. Using a geophone, they were taught to identify and locate these sounds and, with the aid of a chart and an electrical microphone, to calculate the exact distance of the site where the sound originated.

When Woodward returned to Hill 63, he was heartened by the progress the men had made. Major Hill MC, a mining engineer who Bean wrote 'was constantly at them to increase the speed of the work', reported that the Australian companies 'were exception-ally effective, provided they were given some vital task to work off their energies'.[11] Seeming to echo Woodward's disappointment at being given defensive tasks, he wrote: 'listening and pumping was not enough – their keenness and efficiency are too great for nominal defensive [work]' and that it was 'Godsent' that work at Hill 63 had been found for these Australians where they could tunnel to their heart's content into the bowels of the hill.[12]

The massive Hill 63 dugout was completed in just nine weeks, with nearly 200 men employed on its construction, under Woodward's command. Officially named the Wallangatta Dugouts, though popularly known as 'the Catacombs', its entrance was adorned with a cut-out metal kangaroo. Inside there was sleeping accommodation for 1200 men in bunks, which was unique so close to the frontline. The entrance had gas-proof doors, and the entrance tunnel was spacious at nearly two metres square. There was also a good ventilation system and electric lighting throughout, the work no doubt of the 'Alphabet Company'. An impressive and comfortable billet for the troops, it often comes up in soldiers' diaries as having been a welcome and safe refuge from the frontline.

The Wallangatta Dugouts were officially opened to great ceremony by General Sir Herbert Plumer in the presence of a swag of generals and high-ranking staff. It was rumoured that when the Germans observed the many vehicles, officers and men gathering at the nearby Hyde Park Corner intersection, they 'stood to', expecting a British attack.

As the officer in charge of the work, this was a very special day for Oliver Woodward. There was a band and a fanfare of trumpets to welcome General Plumer, and after welcoming the high-ranking guests Woodward escorted them around the new dugout, commending the hard work of his men and pointing out the fine timber lining and overhead lighting system. '[I was] in a world far removed from War . . . but with a thud I returned to reality,' he wrote.[13] For just then, another Australian officer sidled up and whispered in his ear. Unfortunately, but not completely unexpectedly, a group of tunnellers had pinched the rum and were in a sorry state, sprawled out on the proposed tour route ahead. 'Get the blankards out of here – get them out of sight at least,' he whispered back.[14] The officer promptly disappeared into the labyrinth while Woodward quickly devised a new route.

Later he gave the men a good talking to, but took no disciplinary action. 'Had not the Sappers adequately punished themselves? By their own foolish action they had missed the wonderful sight of twenty-odd Generals, a happening which they will regret when they are old men,' he wrote.[15]

BENEATH HILL 60

Woodward's innate understanding of when to punish his men gave them a respect for their commanding officer. Indeed, they appeared to like him. Sapper Frederick Tiffin, who had come to France on the *Ulysses*, made a small wooden box from timber he had salvaged from the destroyed Cloth Hall in Ypres and gave this to Woodward as a sign of his respect. Woodward's family still has it today.

With the generals back in their cars and now safe far away behind the lines, Woodward congratulated his men, shook hands with his fellow officers and poured himself a stiff rum. He felt very proud. As the infantry moved in and dropped their muddy rifles on his clean floorboards, so ended 'one of the most romantic days' of Woodward's life as a soldier.[16]

A major chapter in Oliver Woodward's life had come to a close. And now he would have his chance to prove that he and his men had what it took for offensive mining. For they had already begun their phased takeover of mining operations about ten kilometres to the north, at the notoriously dangerous Hill 60.

156

FIFTEEN

The Move
to Hill 60

In early October 1916, while the finishing touches were being put on the massive dugout at Hill 63, three sections of the 1st Australian Tunnelling Company had gone to the frontline opposite Hill 60. By early November, only Woodward's section remained, and on the 3rd they set out from Ploegsteert Wood sector for the Lille Gate, Ypres. It was dark and quite cold as they departed that night for their posting to Hill 60. Woodward assembled the men at White Gates, a crossroads behind the northeast of Hill 63, to await the motor transport that would take them and their pile of kitbags and cases towards the front. Woodward had long known of Hill 60 and its reputation: the mining and underground fighting and the blood that had been shed for this tiny square of high ground. Among the tunnellers, the talk was of survival, tunnel depths, escape and rescue, of 'Proto' equipment, grey-blue clay and running sand, and of death, the enemy and the legend of Hill 60. For Woodward, the task was a source of pride and honour rather than of fear and apprehension. To him, the Ypres salient:

. . . was so rich in tradition of the deeds of the British Army that all service in this area was regarded as setting the mark of efficiency on troops selected for duty there. In the case of tunnelling companies, service in the Salient carried an additional honour in that Hill 60 marked the spot of the first large scale British Mining activity. Thus not only were we called upon to acquit ourselves like men of the British Army, but in addition had jealously to guard the honour of our particular Branch of the Service – that of Tunnelling.[1]

Sitting in the back of canvas-covered trucks, the men jolted north parallel to the front. The trucks travelled without lights, bumping along the rutted road in convoy, their hard rubber tyres finding every pothole and broken log on the corduroy road. But even at this snail's pace, the cold wind lifted the mud-splattered tarpaulin and froze the line of men sitting huddled on the hard wooden benches. And out to the east, there was the unending flash of artillery, flares and crashing explosions. Welcome to the salient.

A lurch and the crunch of gears and they had arrived at Kruis-straat, on the outskirts of Ypres. Here Woodward was informed that he was to be commanded by Captain Avery, and the tour of duty on Hill 60 would be four days in the line and four days' rest in camp.

From here there were no roads fit for motorised transport, and even if there had been, the noisy, smoky vehicles would have drawn fire from the Germans on Hill 60. The men grabbed their rifles with frozen fingers, and in single file they tramped off into the night, the blackness consuming them and the sound of their boots muffled by the mud. On they went towards the German line, first passing the dangerous Shrapnel Corner, just south of

Ypres and the Lille Gate, where other Australian tunnellers had been busy building a dugout for the AIF in the rampart walls. From here, they took to the duckboard track, past Transport Farm, then along the Fosse Way communication trench, an approach to the front that was regularly shelled.

Ahead, flares lit the sky and the sounds of battle could be clearly heard in the cold night air. German shells began landing in Larch Wood, a headquarters and casualty clearing area ahead on their route. The men, unfamiliar with the sound of incoming artillery, dived for the mud and lay prone while the shells crashed along the railway embankment and the British support line. It was soon over. Dragging themselves out of the mud, the party moved off, wet and cold. The smoke and dust drifted back, and the pungent smell of explosives came to them, distinct and frightening.

The men passed through Larch Wood and quietly eased into the frontline at 10.30 pm, two hours after they had set out. Less than 100 metres away was the German frontline. The enemy had dug down through the bodies of the French, British and their own dead to re-form the shattered ground into something resembling defensive trenches. In the Allied trenches, sandbags lined the parapet and in places timber supported the muddy, weeping walls. Duckboards raised the trench floor above the water level and the fire step had been made stable. They could sense immediately that life was wretched, tense and desperate.

Woodward hardly had time to drop his bag and rifle before he was informed that his first shift would start at 1 am. He took instructions on what the well-dressed, fashionable mining officer was to wear in the frontline, and he was not impressed. Over his drab, baggy Australian woollen uniform trousers he dragged heavy, thigh-high rubber boots and then, to dampen sound, wrapped them in sandbags and tied them at the sides. Next, he took off his Sam Browne belt and replaced it with the khaki web belt he'd been issued, onto which he threaded his holster and pistol. Over his head and onto his chest went his gas mask, in what the training manual called the 'alert' or 'ready' position. And then onto his head went a rolled-up sandbag, an annoying, itchy accessory. 'Even when clothed in the regulation Officers' Uniform I was never sufficiently vain to imagine I might be mistaken for a guardsman, but attired in this rigout I began to develop an inferiority complex,' he wrote.[2]

Woodward related a story of an Australian officer decked out in this assortment of clothes passing along a frontline trench occupied by a British unit. As he went, he called out 'Make way for an officer,' but the Tommies, well used to the humour of Australians, failed to stand aside. The officer reprimanded them and demanded an apology. This they immediately offered, but the Australian officer turned on them again, telling them they had no reasonable excuse not to recognise his rank and 'voice of authority'.[3] And the Australians complained of English arrogance and stupidity. They could find it in their own ranks, too.

Suitably attired, Oliver Woodward set out for his first tour of duty on Hill 60. He and a Canadian officer went down along the tunnel that ran parallel to the British front and stopped off at the 26 listening posts spread out on the Hill 60 side of the railway cutting. At one point Woodward stopped and crouched to use a listener's geophones for the first time since he had been trained in their use at the tunnelling school at Proven. Unlike at the school, which was far behind the lines, noise seemed to come from everywhere. There were the surface explosions of shells, the trench work above, electric motors and the hum of ventilation. He was totally shocked by his inability to interpret the cacophony. To him, it sounded as though the Germans were tunnelling all around him.

The upper level of tunnels beneath Hill 60 comprised the defensive mining system, which was very extensive. From the frontline trench five vertical shafts dropped down to a depth of five metres. From these, parallel tunnels were pushed forward underground, like five fingers stretching towards the German frontline. At 25 metres out, they stopped and a tunnel was dug to connect them, running parallel to the Allied frontline. Along this tunnel five-metre-long tunnels were dug ten metres apart, towards the German frontline. At the end of each, short tunnels were dug to the left and right so that it terminated in a T shape. Listening posts – 26 in total – were established there. On the other side of the railway cutting, opposite the Caterpillar, there were two shafts protecting 100 metres of frontline. Known by the names 'Hooks' and 'Eyes', these two shafts led down to the lateral tunnel, where listening posts were also established at ten-metre intervals. The defensive system was only five metres below ground, so could collapse if struck by a heavy shell. For the tunnellers at this level, death could come from above or below.

Then there was the intermediate level, dug by the Canadians. The entrance to the Berlin Tunnel was in the railway embankment well back from the frontline, in the main support trench known as Bentham Road. The tunnel was an incline that after 150 metres dropped to a depth of about 14 metres. A gallery had been driven to the left, extending for 45 metres, where it branched to the left and right. It was known as the intermediate or 'D' System and the galleries were respectively known as 'D' Left and 'D' Right.[4] From here, the Canadians had broken through into the German workings, captured some sections of the enemy's gallery and destroyed others, particularly communication shafts, with camouflets. This intermediate area was crucial to the whole Hill 60 system as it was hoped that containing the fighting and countermining to this area would lead the enemy to believe that this was the deepest level of the Allied system.

In fact, the real danger for the Germans lay in the deep level, which contained two galleries loaded with the massive mines that would be detonated on the morning of the offensive. These were accessed from the Berlin Tunnel, which, after the intermediate level, continued, the incline finally descending to a depth of nearly 28 metres. Here to the left was a chamber nearly 100 metres long.

Another tunnel had been dug out from the Berlin Sap to the right for 150 metres to a point just past the Caterpillar spoil dump. By August 1916, the Canadians had laid into the Hill 60 gallery high explosives including 21 tonnes of ammonal and 3.5 tonnes of guncotton packed into petrol tins and sealed with pitch. They had completed the tamping of the mine with bags of earth taken from their digging beneath the Caterpillar.

No sooner had they finished this than water burst in, flooding the Berlin Tunnel and cutting off the newly laid charge beneath Hill 60. The tunnel was also filled with carbon monoxide gas from the firing of German camouflets and so a double problem confronted the Canadian tunnellers. Though the Australian Alphabet Company had installed electricity for pumps and ventilation, this equipment soon broke down and hand pumps had to be used. Once the Canadians had pumped the water out they were then able to check their explosives. Fortunately the petrol tins and pitch had kept out the moisture, and the explosives were sound and undamaged. They added more ammonal, and nine firing circuits with more than 60 detonators spread amongst the charges. By October 1916 they had also packed the gallery under the Caterpillar with 32 tonnes of ammonal and tamped it, ready for firing.

During this work, they broke into a German gallery, which they found extended well behind the British lines but had been abandoned. They quickly incorporated this into their new system. However, soon afterwards the Germans were heard pushing a tunnel forward very close to the already tamped Hill 60 mine, and GHQ, knowing the work and the fatalities involved in its construction, gave permission to fire the mine if necessary. In the end it was not necessary, as the Canadians cleverly diverted the German effort by selectively countermining their galleries and putting them off their work.

At 7 am on 4 November 1916, as Oliver Woodward staggered exhausted from his first shift underground at Hill 60, he was familiar with the tunnels, the listening posts and the German work in the area, and he had a new appreciation for the men who formed the first line of underground defence, the men of the uncelebrated and largely ignored listening service.

As the cold, black night gave way to the weak rays of the dawn sunlight, Oliver Woodward slipped into his dugout and lay down. Although it was only a six-hour shift, it had been an arduous one, especially after the exhausting trek up to the line. He had secured two extra blankets, good Australian wool ones, too, but these were not enough to keep out the cold. He felt the chill along the length of his spine and shivered as he drew up his knees and pulled the itchy woollen blankets over his head. He had done this in Tenterfield as a kid when the winter temperatures had dropped and the frost had covered the front lawn. How he wished he were in Tenterfield. He thought of his family and the boys he had grown up with – so many of them here in Belgium now, shivering just like him.

✕

Until they were called upon to explode them, the Australians' main task was to protect the two great mines from German countermining and deterioration in the damp underground chambers. They soon found that there were major drainage and ventilation problems to resolve, and the ageing equipment needed immediate attention. Water continually seeped into the tunnels and threatened to ruin the explosives, and they would need to shore up and repair tunnels to prevent collapse. In time, the detonators and the explosive material would become wet and less likely to explode. And the longer the offensive was put off, the greater the chance that the Germans would discover the galleries and dismantle the charges. Over the coming months they would test and re-test the mines to make sure that the explosives and the electrical leads remained viable. And they would construct dugouts and other defensive works. There was much to do.

The first thing the Australians set to work on when they arrived was sorting out the Berlin Tunnel, which had been fought over, countermined and flooded, causing the Canadian tunnellers continual problems. It was hated by the infantry work parties who dragged endless amounts of spoil from it and worked pumps to drain an equally endless amount of water. The Australians began by sinking a new metal-lined shaft 30 metres deep and two metres in diameter. Then, from the bottom, a further gallery was driven down an incline. This allowed the water to drain into a sump, from which it could be pumped out with an electric pump. They started a new system of tunnels at the bottom of the shaft. The intention was to drive a gallery under the German strongpoint known as the 'Snout', which was 425 metres away to the left of Hill 60, but the works were not completed by June 1917.

Woodward soon understood the trenches and defensive

systems above ground at Hill 60 and the direction a German attack was likely to come from. This was crucial as the tunnellers were often called upon to take up their rifles and line the parapet to repulse attacks. He began surveying and mapping the underground workings, logging the daily reports from the listening posts and plotting the slow advance of the enemy beneath them. He was keen to learn what the Royal Engineers Listening Unit was picking up from eavesdropping on German telephone traffic. They were using a contraption of valves and tubes resembling a wireless set, attached to a copper plate in the Hill 60 mining system to pick up and then amplify calls being made on the German telephone system.

On 9 November 1916, the Canadians officially handed over the nightmare of Hill 60 to the 1st Australian Tunnelling Company. 'From then on the care of the Hill 60 Mining System rested on our shoulders,' Woodward wrote.[5]

The next day, the Germans blew a camouflet in one of the Australians' dugouts, killing two men sleeping there and gassing three officers. On hearing this, Woodward, who was at the time having his four days off in billets near the Lille Gate, hurried back to Hill 60 after dark with another officer. He wanted to assess the damage personally, and so upon arriving at the Hill he descended to the intermediate galleries, the most dangerous as here the German gallery was within two to three metres of some of the listening posts. 'I have listened . . . as immovable as a piece of statuary and have been equally as cold, from fear,' he wrote. 'It is rather thrilling when one is stalking, but decidedly nerve-racking when one is being stalked.'[6]

Life was no less dangerous in the trenches above. The Germans kept up a relentless shelling of the Allied front near Hill 60. In mid-November, two good men, Sergeant William Ruddick and

Corporal Mudie, were killed, and Sergeant Thomson had his right hand blown off. Thomson, a miner from Toowoomba in his early thirties, spent six months in hospital in England recuperating. Just a week before the Messines offensive, he was sent to work at the massive Allied Base Camp at Étaples near Boulogne but soon deserted and headed to the front. He was arrested by the military police at Poperinghe, 11 kilometres behind the line, but his commanding officer needed men like Thomson – even missing one hand – so he was released and returned to the line, finishing the war as a lieutenant and being awarded a Military Cross and a Distinguished Conduct Medal.

Oliver Woodward was appointed to be the officer in charge of the Hill 60 and Caterpillar mining operations on 24 November 1916, and given a temporary promotion to the rank of captain, backdated to 23 October. In November Woodward was made the commanding officer of No. 2 Section. This was an enormous responsibility, and for him an amazing honour given the strategic importance of this position on the British front. It was now his duty to protect and maintain the massive mines. He was proud that his Australian tunnellers had been considered worthy of the task, and he was sure they would be proud of him back in Tenterfield.

On the morning of 26 November, he had just woken when a breathless tunneller knocked on his dugout and, pulling the sandbag curtain back, explained that a distinct current of fresh air was blowing through his listening post. This was serious. Had the Germans broken through at some point, perhaps unknowingly damaging the timber lining boards? Were they at this moment preparing a camouflet? Forgetting his boots but grabbing his pistol, Woodward leapt out of bed and, keeping his head low, ran along the trench and dived through the shaft entrance. He hurried down the sloping gallery, his stockinged feet barely making a sound. Along the lateral he went, careful not to stumble, and found himself at Listening Post No. 25 beneath the frontline.

Crawling into the tiny space, he made his way to the face. Yes, there was a draft, a faint zephyr, but it was difficult to determine exactly where it was coming from. Slowly, he removed a section of the timbered panelling and there ahead was the ominous black hole that indicated another shaft. Immediately the draft increased. Probing ahead very carefully, he pulled back the boards and poked his head through the jagged hole. Beyond it was black.

Taking his small battery torch from his pocket, he flicked on the light, half-expecting to see a blond German in the beam with a pistol aimed at his head. Instead, he discovered a shaft and a ladder, both in a good state of repair although the upper section had been blown in and collapsed. This, he realised, was one of the disused tunnels that honeycombed the hill, probably dug long ago, possibly by the French and forgotten, as it was not marked on any of the mine maps left by the Canadians. The panic over, he headed back to the surface to get on with a new day.

Late in November, Woodward observed a German spoil dump building up, a clear sign of new German mining activity. Sure enough, the daily reports from the listening posts started to include references such as 'Noises heard. Nothing definite'. He asked that aerial photographs be taken over the German front-line, a risky procedure for the pilot and photographer. These confirmed new shafts had been dug, and he added these positions to the mine maps he was building up.

On 1 December, a listener reported: 'Berlin sap. Sounds of enemy working picked up by geophone. These sounds were also heard again today [but] it has not been possible to ascertain in the exact direction.'[7] Then, five days later: 'Main Berlin – distinctive noises have been heard to the left and ahead but some distance above.'[8] Two days later in the Caterpillar sap: 'Distinct sounds of hammering confirmed by officer. Today sounds of continuous face work estimated to be between 30' and 40' distant.'[9]

The concern was that if the Germans continued their vertical shaft, they might well discover the Hill 60 mine or begin countermining that might prematurely fire the huge charge. Should the Australians begin countermining, there was a fear that this, too, might trigger the mine and totally waste the effort and the lives of the many miners who had worked on it.

'Captain Woodward and Lieutenant Clinton estimate within [six metres] . . . sounds of digging and falling earth most audible and becoming much closer in last 36 hours,' it was reported on 15 December. On the same day, 'The enemy bombarded this morning for two and a half hours, again this evening doing considerable damage to trenches . . . It was thought the enemy would attack this evening. All tunnellers stood to for two hours.'[10]

Woodward had decided to drive a gallery to intersect the German shaft and blow it when the time was right. By 17 December the Australian tunnel was just a metre directly beneath the German workings. Two days later, orders came for Woodward to act.

Please take immediate steps to have D-right driven with object of entering German galleries. Our object is to capture as much of his gallery as we can. Every precaution must be taken for the possible influx of gas. Should you hear the enemy and have any reason to believe he is loading, you will blow with all possible speed if he is within striking distance of your mobile charges. Should this be done, immediately report the matter to me.

Signed J. Douglas —— (name indecipherable)[11]

The following day, the Germans could be heard with the naked ear they were so close. A charge of 1200 kilograms of ammonal was quietly laid in the Australian gallery.

The tunnellers warned the British infantry along this section of the front that the countermining might ignite the massive charge under Hill 60, and the decision was made to fire the camouflet at 2 am on 19 December 1916. At 4 the morning before, Woodward's men completed the placement of the detonators and ran out the firing wire. The German tunnellers were so close – just a metre or two separated them – that their tunnelling caused flakes of clay to begin falling from the roof. The Australians had to quickly cover their tins of ammonal with sandbags to muffle the noise. 'It seemed that at any moment we could expect the bottom of the enemy's shaft to fall away and precipitate earth and enemy on top of us,' Woodward wrote.[12]

By 6 pm on the 18th the electrical leads had been stretched back to the firing position and the tamping had begun. Tamping was the process of blocking the tunnel leading to the gallery full of explosives with tightly packed bags of earth. When a charge is ignited, the explosive force follows the path of least resistance, so if the tunnel were not blocked, the force of the explosion would be lost – it would not explode upward but be dispersed sideways underground. To counter this, after the charge was laid and the detonators put into place, the tunnel leading to the gallery full of explosives was tightly filled with bags of earth for a distance of up to 20 metres. This ensured that the tunnel was stronger than the earth above it, forcing the explosive's energy upwards. A small air gap was left as air gaps help withstand the shock of an explosion, and more tightly packed bags of earth were placed in the tunnel for a distance of up to ten metres.

It was a hard and dangerous job, and they had finished by midnight. At 1.30 am on the 19th, the men were withdrawn from the frontline saps and the infantry took up their positions. All was quiet. Woodward had left a remote microphone with the charge, and the sounds of German activity came clearly up the line to him. The Germans were still at work.

At 2 am, the mine was fired. A huge tongue of flame leapt from the shaft of the German workings and for a brief moment lit up the sky. Gas spewed through the fractured earth, and everyone in the immediate area was killed. So successful was the camouflet that it was not until March the following year that German mining activity was heard in this area. The only damage the Australians suffered was to one of the sets of firing leads on the big mine, which had been broken.

The Germans fought back the following day by firing a charge at a depth of six metres above and to the right of D-Right. A six-metre-long section of the Australian gallery was smashed in, and a listening post was slightly damaged.

Woodward ordered that the tamping be removed from the big mine so that the leads could be repaired. Rather than re-tamp the mine, a listening post was established close to it so that the German tunnelling could be monitored from this important position. This quickly showed that the Germans had not been directly targeting the big mine, but rather desperately searching for the Allies' gallery. Still, it was a cause of great concern and anxiety, because if the Germans began indiscriminately firing camouflets they might inadvertently trigger the mine or cause enough damage to render it ineffective.

Water seeping into tunnels was an eternal problem. In January 1917 flooding caused work had to stop and the water had to be pumped out. Continual German bombardments caused delays in the men's work underground, as they would be commanded to 'stand to' for hours at a stretch. Work was frustrating, but water was not their main concern: the Germans were getting closer.

SIXTEEN

A Month Today

The French, under their new commander General Robert Nivelle, commenced a massive offensive on 16 April 1917 across the River Aisne, on 65 kilometres of the front between Soissons and Reims in Champagne to the east of Paris. This was about 120 kilometres from Amiens, southeast of the area of Australian operations from mid-1916 through to August 1918. Nivelle rashly promised it would end the war, convinced that within 48 hours he would have broken through the German line and would be free to go on to the Rhine.[1] To assist the French and to divert German attention from the south, on 9 April 1917 three British armies had committed to an attack at Arras, a large provincial town on the frontline just to the south of Vimy Ridge and between Amiens and Lille.

The offensive at Arras gained traction, especially when the Canadians took Vimy Ridge, but General Nivelle's grand plans quickly ground to a halt in the south. From the start, things began to go wrong. The Germans, well aware of the French battle plan, quickly counterattacked along a 65-kilometre front from Soissons to Reims. Within four days, the French had 120,000 casualties.

Sixty-eight of the French army's 112 divisions mutinied in response.

The Eastern Front was collapsing. The Russian army, dispirited, poorly led and hungry, was feeling the effects of the festering Bolshevik revolution. With shortages of ammunition and material, they fell back in disarray. This eased the pressure on the Germans, allowing divisions from the Eastern Front to be spirited off to the Western Front, where they reinforced the spreading *Siegfried Line*, known to the Allies as the Hindenburg Line. The United States had declared war on Germany on 6 April, but it would be some time before their troops were at the frontline. The British position looked decidedly grim.

Haig now sought to mount a battle with immediate and substantial results, and where better than the Messines Ridge? British Prime Minister Lloyd George had subjugated Haig to French command, so he could not dictate Allied strategy for the year 1917. But after the failure of the French offensives, and having been publicly discredited for his support of General Nivelle and the transfer of field command from Haig to the French, Lloyd George now felt he had little choice but to give Haig a go. And so the Battle of Messines was given the official nod. On 7 May 1917, Haig summoned General Plumer to dinner and asked how much time he needed to launch the Messines offensive. Without hesitation, Plumer responded, 'A month today', and so the date was set: 7 June 1917.

✕

At Hill 60, Woodward's greatest concern was that the Germans were persisting in their search for the Allies' deep mines. In early April, while supervising tamping of the Hill 60 mine, Lieutenant R. B. Hinder took time to close down the Australian work and listen through his geophone. He picked up the sound of a German winch working in a nearby shaft.

At a listening post on 4 April, Corporal Snedden heard a lot of German activity, and fearing that they were about to blow a camouflet he withdrew the listeners from the tunnel. When nothing happened, Snedden returned alone. 'He had just about reached the post when the enemy fired the mine, wrecking the gallery. We at once set out to re-open the gallery and after several hours' work reached the body of the Corporal,' Woodward wrote. That Corporal Snedden's son was a member of the same unit must have made his death especially difficult for everyone. Woodward continued: 'Next morning in almost similar circumstances Corporal O'Dea lost his life in the Hooks & Eyes System. In the War the loss of a couple of lives here and there seemed to count for little, but to the individual Company this slow but regular wastage did not pass unnoticed.'[2]

On 9 April, the Germans raided the British trenches in the hope of finding and destroying the mine shafts and the workings. Though the German raiding party roamed 200 metres behind the lines for more than an hour, they did not find the mine entrances. They did, however, do a great deal of damage, blowing in the entrances to shallow subways constructed for the safe passage of the infantry and capturing five Australian sappers. It was feared they might talk and give the location of the mine away. In attempting to repulse the attack, ten men of Woodward's company were killed, and the infantry lost a further 23. The next day, the bodies of 33 of the enemy were found in the lines.

Needing to keep close track of the advancing German tunnellers, some of the tamping was removed and experienced listeners were placed along the galleries leading to both the Hill 60 and Caterpillar mines. The listeners had 'got the wind up' and begun hearing strange noises, which they were quick to report. The sound of the winch could still be heard. It seemed that it was most likely bailing water. In the words of Corporal Gough: 'It was too constant to be hauling dirt, for they could not break enough to keep it up.' From the way it creaked, it appeared that the Germans were using very old equipment.[3]

With tension rising, on 17 April two very experienced listeners were brought in: Captain Pollock, Professor of Physics at the University of Sydney and at the time in charge of the Army School of Mines, and Lieutenant R. F. Clarke MC of the Canadian 3rd Tunnelling Company, who was a surveyor in civilian life. After carefully listening from a number of listening posts and triangulating their readings, they determined that the Germans were 80 feet away and getting closer. To counter the Germans, a gallery was begun and pushed out from the Hill 60 tunnel, but before this had gone far, underground fighting started in the intermediate level of tunnels. The Australians blew two camouflets on 18 April, and the Germans responded with one on the 19th and two the following day. On 24 April a listener heard enemy footsteps so close that he extinguished his light and prepared to defend his post. The footsteps passed less than two metres over his head.

×

When the primer for an explosive charge was being prepared, the standard procedure was to test the electrical continuity of the detonators before inserting them. But in late April, when a primer was being assembled in the headquarters dugout, Captain W. P. Avery, a mining engineer from Brisbane, inserted electrical detonators into a 50-pound box of guncotton, having forgotten to test them first. 'Evidently it was decided to test the detonators while they were in the primer, and by a thousand to one chance there must have been a supersensitive detonator which exploded when the testing current was put through the circuit,' wrote Wood-ward.[4] The explosion destroyed the dugout and killed Captain Avery, Lieutenants Tandy and Evans and eight sappers.

Lieutenant McBride took on the very macabre task of sieving through the debris to recover every possible piece of the men killed in the dugout, earning the men's respect for his dedication and compassion for his mates. He was later awarded the Military Cross for conspicuous gallantry and devotion to duty, organising the defence of a deep-mine system and arranging parties to rescue several men who had been gassed. His prompt action undoubtedly saved many lives.[5]

By early May, the first signs that an offensive was being planned were becoming apparent. In the towns well behind the line there was a steady build-up of troops as the infantry began to assemble. Trains brought up supplies, ammunition, and the new Mark IV tanks that were to take part in the attack. Great piles of artillery shells came up with the artillery units. The roads were clogged with slow-moving gun limbers, horse-drawn wagons and lumbering tractors drawing heavy siege artillery and battalion upon battalion of marching men, all heading for the front. Vast camps were established to feed and rest the men. Well-camouflaged artillery seemingly disappeared into their firing positions.

No sooner had the batteries of artillery been dug in and covered than they began to fire. To confuse the Germans, the Allies mounted long, harassing bombardments upon their defences and far into their back areas. Without warning, every gun in the British line would open up, sometimes just for 15 minutes or perhaps an hour, but each time the Germans would expect the anticipated and long-awaited attack and rush to their parapet and 'stand to'. They were continually strafed, gassed, annoyed and kept from their sleep to break down their resistance and shatter their nerves. Just when the tension was at its height, the guns would fall silent, and the men would file back to the dugouts and wait for the next Allied false alarm.

Between these organised bombardments, individual field guns would randomly fire on the German trench lines, mobile cook houses, assembly points, artillery positions and supply lines. With the sophisticated listening equipment the British had deployed, they could accurately pinpoint the location of German batteries, which they systematically shelled. These guns could range deep into the German rear areas, striking targets with surprising accuracy 15 kilometres away.

The Allies had finally grasped that it was pointless taking the first few lines of German trenches in an offensive without taking out their artillery beforehand. Left operational, the German artillery would simply blast them out of their newly won positions. The Germans knew the ground, especially their own, and to the metre. Their guns could range on any newly won trench line with accuracy and consistency. They could target a tree, follow a vehicle or stalk a single man. In the lead-up to the firing of the mines, the Allies identified and neutralised their targets well, and it is believed they destroyed 90 per cent of the German artillery before zero hour.

For the Germans there was no let-up. They knew the offensive would be coming soon, supported by waves of infantry, massed artillery, tanks and the possibility of underground mines. Though they had been reassured about their defensive position and the fact that the British had not succeeded in mining their strong-points, they were deeply fearful that the British had something big up their sleeve.

×

At night, Woodward continued his long patrols underground, visiting the listening posts, checking the men's quiet digging and testing the charges. He had completely replaced and re-designed the firing process, complete with a range of back-up procedures. The success or failure of the mines all came down to these thin firing wires that carried the all-important electrical current. Down through the dark shafts and passages these leads snaked, some protected in heavy metal tubing, along wet, damp galleries, past small cave-ins and bulging timbered walls, under bags of tamping and on into the vast chambers filled with explosives. Tin upon tin, carefully sealed, caked in pitch and packed high to the muddy, dripping ceiling. It was a frightening sight: two vast explosive charges just waiting for a trickle of electricity, that small surge of power that would fire tiny detonators – small explosive charges – that were packed in with the ammonal and guncotton. Their ignition would set off the main explosive charge. The Hill 60 mine had been in position for nine months, the Caterpillar mine for seven, and their deteriorating condition was a great source of concern. Rain fell and drained into the tunnels, and rust had started its callous and corrosive work.

Woodward was leaving nothing to chance: he made sure there were two firing systems should one system fail. One was a Service Exploder, a small dynamo that generated an electrical current when the handle was depressed. The second system relied on the electrical current from a 500-volt generator normally used to power lighting. Woodward thought it unwise to depend solely on this form of power as the generator was located in a dugout; with just over three metres of overhead cover, a chance shell might take out the system at the last minute.

He was cautious, too, about providing back-up electrical leads: every mine had three sets of electrical circuits, each with five detonators in series. In the lead-up to the exploding of the mines, these leads were connected to a central switchboard in a dugout. This allowed the leads to be tested frequently to ensure that they had not deteriorated or been severed by the Germans.

There were two types of tests that Woodward ordered. A continuity test was used to make sure that the electrical circuit was still intact. A battery and galvanometer – an instrument that detects electrical current – were connected to each set of leads. He would watch and hope that the needle on the galvanometer moved, showing that electrical current was passing from the battery through the leads to the detonators.[6]

It was the second type of test – of the circuit's electrical resistance – that Woodward considered to be of greatest importance. If the enemy had broken into the tunnel, cut a set of leads and then joined the ends, the continuity test would still show a current passing through the circuit, even though it might not be enough to fire the detonators. The material that makes up electrical leads puts up a certain amount of resistance to an electrical current and converts it into heat. This can be measured using an instrument called a Wheatstone Bridge. After laying the leads, Woodward had calculated the exact resistance there should be in each circuit, based on the length of the leads and the resistance per foot of those leads. This figure became the benchmark against which the results of ongoing regular resistance tests were compared. If the resistance was measured with a Wheatstone Bridge and found to match the benchmark, the current passing through the circuit was deemed to be correct.

Even the resistance test was not foolproof, though. If the enemy discovered the leads of a mine, they could cut them, then use a Wheatstone Bridge to measure the resistance in the section of the circuit leading from that point to the mine. If they then attached the Wheatstone Bridge to the part of the circuit leading back to the Allies' testing equipment, the resistance and continuity tests would appear fine, though in fact the leads weren't connected to the mine at all. When the plunger was pushed down, there would be no detonation. Woodward was well aware of the possibility of this kind of sabotage because the Allies had themselves done it to the Germans. He wrote in his diary: 'One of the Royal Engineer Tunnelling Company actually performed the feat, and removed the whole of the enemy's mine.'[7] Woodward had resistance tests carried out almost continuously, in the hope that if the Germans made such an attempt they could catch them in the act.

Woodward was worried because he was getting a weaker and weaker signal each time he ran the tests. The circuits were deteriorating. He worried about how to prevent moisture from damaging the explosives, and about the structural strength of the mine supports that he and his men had installed to replace the saturated, splintered timber they had inherited from the Canadians. They had made repairs in the tunnels, improved the tamping and maintained the firing wires – he hoped it was enough. Tension was daily increasing as the listeners could hear the German mining works closing in on the Hill 60 mine. There was also the possibility that the Germans were driving a tunnel between the two charges on Hill 60 and the Caterpillar. These thoughts raced through his mind, tormented him and kept him awake at night. The announcement of zero hour must be soon, he hoped.

In mid-May 1917, he was summoned to headquarters at the request of his Commanding Officer, and was informed that it was to be his job to fire the two mines when the offensive began. He recorded in his diary:

> I felt that upon my shoulders had been placed a heavy
> responsibility . . . Any slip on my part would endanger the success
> of the attack, and increase the loss of human lives. Thus my earnest
> hope was that I would prove equal to the task, not to satisfy
> personal desires but rather that no failure on my part would render
> futile the work of thousands of comrades-in-arms of tunnelling
> companies.[8]

Woodward's diary makes no mention of the Germans who would lose their lives when he detonated the mine and their front-line disappeared. His only thoughts were that he must get it right, and the enormous responsibility that the job entailed.

On 21 May 1917, Woodward was sitting alone in his dugout when the hessian curtain was suddenly drawn back and an unfamiliar face appeared. It was Captain Worlledge of the Royal Engineers who had come to go over the plans for the firing of the mines.

For the next four hours, the captain made an exhaustive examination of the Hill 60 system. He pored over the mine maps, asked about the condition of the charges and the tamping, tested the leads and discussed all details of the method of firing. He discussed the German threat and Woodward told of his concern for the German gallery that was working its way towards the Hill 60 mine. Then Captain Worlledge gathered up his leather shoulder bag and walked to the dugout entrance. He said he was happy with the progress and the work of the Australians and approved all of the plans Woodward had put forward for the firing. He also indicated that the long-awaited attack was close at hand, so attention to detail and continual testing and checking were now most important. With a slight bow of his head, he shook Woodward's hand, turned on his heel and disappeared with a clatter along the duckboarded trench.

A wave of mixed emotions flooded over Woodward. The realisation that the attack was imminent 'seemed to unconsciously add to the strain', he wrote.[9]

Woodward's greatest concern remained the reports of German mining towards the deep gallery housing the massive ammonal charge. On 8 May he had crawled to the listening post in a section of the untamped gallery and there confirmed for himself the sound of German work. He ordered a camouflet of 1600 pounds of ammonal to be prepared and placed the following day, but told the men that on no condition must it be fired without his orders. It could well detonate the main mine, with terrible consequences. Unless the Germans actually broke into the Allies' shaft, he would accept the risk of them finding the mine and let the enemy work on.[10]

Then on 25 May, the Germans fired a camouflet directly above the Hill 60 gallery, in a position that was reported as 'dangerously close' to the massive mine. The force of the explosion collapsed a gallery in which two listeners had been stationed, cutting them off in a small length of tunnel. It was feared that both men had died and been buried, but on the day following the explosion, while clearing debris from the blast, Sapper G. Goodwin heard faint tapping from ten metres away. The sound was from Sapper E. W. Earl, a labourer from Geelong in Victoria, who had been quietly and calmly writing to his mother and composing his will. Goodwin could also hear the tapping by the other entombed listener, Sapper G. Simpson from Chatswood, New South Wales. After giving his position and knowing the rescuers were on their way, Earl lay silently, not wishing to draw German attention to his location or the position of the mine.[11]

It took two days for a rescue party, under Sergeant H. Fraser, to reach Earl and Simpson. Earl handed the rescuers a written report of what he had observed and heard of the Germans' activity. By the time Earl was rescued, it was too late. He was suffering from asphyxiation and died two months later. The German work became more cautious afterwards, so it seemed they had heard the trapped men's signals or the sounds made by the rescuers.

✕

From the high ground near Ypres, down along the Messines Ridge and south to Factory Farm, near Ploegsteert Wood, the Germans had watched the build-up of men, artillery and supplies behind the British lines. There was no doubt that a major offensive was coming. They were confident in their defensive plans but were very concerned about the possibility of a mine attack. The British mines blown at St Eloi in the spring of 1916 and the massive mines used on the Somme had made a lasting impression. After the war, Oberstleutnant Otto Füsslein, the commander of the miners for the German 4th Army along the Messines front, recalled 'the sudden, fearsome shuddering of the earth, the mightily towering black cloud from which spouted tongues of flame, the rain of earth, timber, iron and shattered human bodies which poured out of the cloud over the surrounding area and the four blackened, yawning craters strewn with corpses open to the heavens where the German positions were'.[12]

They suspected that a number of mines would be involved in the Allied attack, but they were confused by the fact that there was now very little Allied tunnelling activity. They decided they should fire some large mines themselves to see if they could disrupt the Allied preparations. And so they organised the 'Landslide' countermining operation along the Messines–Wytschaete salient. The German plan was to blow large charges in their own workings in the hope of collapsing Allied tunnels and blocking Allied shafts. They would also destroy the entrances and shafts of their own tunnels so that the Allies could not use them against the Germans in the future.

Due to the continual British bombardment of the rear areas, the roads stretching back ten kilometres were churned up and impassable, so the explosives for the operation had to be carried to the German front by men and horses. For the German tunnellers, life was difficult and dangerous. Their generators had been destroyed by the British bombardment, so they had few reliable pumps or ventilation equipment, which made the air in the tunnels heavy and foul. They rushed to lug explosives, tamp and prepare the charges for the operation before the Allied attack, and because of the appalling conditions some exhausted men were collapsing after half an hour's strenuous work. They would stumble, gasping, to the bottom of the entrance shaft, desperate for air and oblivious to the shells falling. Others were far too weak to climb up the metal ladders in the shafts and had to be carried or winched up to the shattered trenches above.

On 31 May, the Germans blew their mines, destroying years of their own work. Staring out across no-man's-land, they saw dust and smoke rising from the British line and knew their charges had crumped in the shafts and entranceways and killed everyone unfortunate enough to be working close by underground. They destroyed some of the Allied workings, at Peckham and Spanbroekmolen in particular, and forced a frenzied effort upon the Allied tunnellers to make good the damage. But the detonation leads and explosive charges for the Allies' mines were encased in heavy steel ducts, which largely protected them. The German attack had, in effect, totally failed.

At points along the front, the Germans considered abandoning their frontline trenches, especially where they knew there were too few German tunnellers available for defensive mining work and where it was anticipated the British mines would be fired. But the German Command forbade this, believing that the Allied mining was only in a couple of possible areas and that the mines, blown in front of their lines, would give due warning that an attack was under way.

✕

It was Saturday 2 June 1917 and Woodward was in billets in the Ypres rampart dugouts when he was pulled aside by his CO and told in confidence the date of the attack: 7 June. It was less than a week away, and there was still much to do. He gathered together the other men who would make up the firing party – Lieutenant Royle, Lieutenant Bowry and Sergeant Wilson – and 40 sappers. Soon afterwards, they were all tramping back along the now-familiar duckboards towards Larch Wood as the sky ahead lit up with the arc of flares and the twinkle of the guns all along the front. He told none of the men the news to which he was so recently privy.

Reaching Hill 60 at 10.30 pm, he immediately directed that work begin on reinforcing the dugout and mining system behind the lines. He wanted to ensure that they could withstand the force of the explosion. This meant adding extra reinforcing and increasing the thickness of the head cover over the firing position by piling on more dirt and sandbags. Nothing could be left to chance, and nothing could be taken for granted.

Now for him it was a game of waiting and hoping that the continual testing would show the mines were ready and that the Germans would not reach the Hill 60 gallery before zero hour. The Germans were just metres from the mine and there was nothing that could be done – certainly a camouflet could not be fired so close to the massive charge. And so he had to wait. With the consequences of failure unimaginable, it was to be a very tense time.

SEVENTEEN
The Days Before

Along ten kilometres of Allied frontline, men were on the brink of nervous collapse. As zero hour approached, they tested the leads more frequently. Mining officers and tunnellers were on edge, as less and less current registered on their test equipment. Each gradation represented a small but significant weakening in the power flowing down the firing wires to the charge. Would there be enough electrical charge getting through to fire the detonators and explode the mines? It was touch and go.

Many of the mines had been in position for over a year and their deteriorating condition was a great source of concern. Rain filled the tunnels now the pumps had fallen silent. Moisture was a relentless, determined and destructive enemy threatening to destroy the work of many men. All their effort, all their pain and blood – and now they simply had to wait. Nerves were frayed to breaking point.

Above ground, the preparations continued. The infantry moved towards the front – endless lines of marching men with heavy packs, basic webbing bulging with small-arms ammunition and grenades, slung rifles, entrenching tools and their gas masks sitting ready across their chests. They had little idea of the battle ahead. All frontline troops fear enemy mines but as they trudged along the muddy Belgian and French lanes that led to the front, the men could little suspect that massive charges lay silent beneath the earth and were about to wreak damage and carnage on the enemy before they sprang across their tapes and headed into no-man's-land in a few days' time.

Three kilometres south of Hill 60, the Canadians at St Eloi were waiting anxiously. On 28 May they had just finished laying the final charges and tamping their mine, the deepest of all the 19 mines at nearly 40 metres, and holding the largest charge of nearly 44 tonnes of ammonal. Unlike Woodward's elaborate firing system, the Canadians had just two circuits, and rather than a comfortable, protected, light-proof firing position, their firing party was to take up a position in a dugout or open trench. They had been under way on a gallery that would have allowed them to fire another mine, but needing another two to three weeks to complete the work, had abandoned this plan.

South of the Canadians, the 250th Tunnelling Company of the Royal Engineers had seven mines to keep operational and ready for firing. Their commanding officer, Major Cecil Cropper, was credited with digging the first deep mines into the blue clay, and he had come up with an idea similar to Norton-Griffiths': to blow open the Messines Ridge with a series of mines. But in late December 1916 he had been ordered to hospital suffering from exhaustion and German measles, and unfortunately could not return to Messines to watch the results of all his hard work on the northern three mines at Hollandscheschuur.

The 250th Tunnelling Company was responsible for three other mine sites. At Petit Bois, the second-longest tunnel stretched well under the German lines to two relatively small mines. At Maedelstede Farm work had fallen far behind schedule, and one shaft was abandoned when it was realised it could not reach its objective in time. All effort went into the main Maedelstede tunnel. When it reached a length of 500 metres, it was loaded with over 40 tonnes of ammonal and 1.8 tonnes of guncotton, tamped, tested and made ready for firing just one day ahead of zero hour. It was the second-largest charge and was to be fired by electrical current from a generator normally used to power lighting. At the Peckham mine, the nuggety Welshman Haydn Rees had battled cave-ins, camouflets, wet sand and a muddy slime that seeped into his gallery and threatened to close his mine. He had laboured to build diversionary tunnels around the cave-ins and re-wired the charges himself.[1]

At Spanbroekmolen, a feverish, incautious group of miners of the 171st Company were pushing relentlessly forward, now knowing the date of the attack and not caring much if the Germans heard them or not. In early March 1917, a German camouflet had rocked the main gallery, severed the firing wires and completely collapsed the tunnel. It had taken nearly two months to drive another parallel shaft; along the way, the tunnellers struck pockets of deadly gas, and three died of asphyxiation. In a race against the clock, Captain Henry Hudspeth dug a diversionary tunnel around the collapse. This allowed the men to attach a half-tonne dynamite priming charge to the original charge of nearly 40 tonnes of ammonal, and run new firing wires back. This was frantic work, the tunnellers sweating and straining and ever fearing another German camouflet. The massive charge was ready just hours before zero hour.[2] Hudspeth was so concerned the mine would not fire that he quickly scribbled a note to his Divisional Headquarters, explaining the problem and asking that the commander of the attacking troops be notified.

He paced about, fearful that the waves of men throwing themselves against the German line in the days ahead might not have the advantage of his massive, destructive mine.

At Kruisstraat, 700 metres away, Captain Henry Hudspeth had other mines ready. He had dug a tunnel over 650 metres long, well behind the German frontline and as far as the German third line of trenches. Here he had laid three mines along the gallery with a total of 50 tonnes of ammonal. These were connected to a generator used for lighting, and were ready for firing. Hudspeth had suggested that these mines be slightly delayed so that the Germans fleeing from the frontline trenches would be caught in his blast. He was refused permission on the grounds that the delayed firing might cause casualties for the advancing infantry, hit either by the shock of the explosion or falling debris from the blast.

Hudspeth also had a mine at Ontario Farm, directly west of the village of Messines. Here, in the second-deepest mine, 32 metres below the surface, he had problems with liquid sand and was rushing to complete the work. This was a crucial mine in the attack on the ridge line, and he knew full well his responsibility and place in the grand plan. By early June, he had excavated the chamber and by the end of the following week, just one day before the attack, had placed 27 tonnes of ammonal and tamped the charge. The firing leads were only run out a few hours before zero hour, connected to the two exploders, and the firing party able to take their places.[3]

At the southern end of the line, in the section furthest away from the Australians on Hill 60, there were still problems. Here the men of the Canadian 3rd Tunnelling Company, who had moved to this sector after handing over Hill 60 to the Australians the previous November, had been working on a series of mines at the very end of the Messines Ridge, just to the east of Ploegsteert Wood. Three days before the start of the attack, a heavy-calibre shell made a direct hit on the entrance of an escape tunnel at Factory Farm. Suddenly, a 20-metre hole opened up and when Lieutenant Hall peered down into the darkness, he could just make out the coils of the firing leads at the bottom, which were being buried in sand cascading from the broken boards that lined the shaft. He scrambled down the ladder and with Sergeant Beer worked frantically to repair the lining boards and recover the buried firing leads.

Originally, Hall and another officer, Lieutenant George Dickson, were to fire their mines from a position at the base of the shaft, but because of the extent of the damage, with zero hour just 12 hours away, their firing position was now to be above ground. This meant running firing leads from the base of the shaft to the surface. There was little time, especially given that there were six separate circuits to connect up. To make matters worse, Dickson was affected by gas. With the help of Sergeant Beer, the men attached the firing wires, fixed them securely to the timber lining and tested the current. Up and down the two men went to check and recheck the circuit, sweating, struggling and frantic until just 20 minutes before zero hour, they made their final test, and waited.

On the night before the attack, General Plumer is said to have remarked to his staff: 'Gentlemen, we may not make history tomorrow, but we shall certainly change the geography.' How right he would prove to be.

✕

Oliver Woodward was nervous. He had kept the tunnel open through to the gallery that contained the ammonal charge, as he was anxious that the approaching Germans might break into it, and he wanted to be able to monitor their movement and attack if needed. On 3 June, in preparation for the firing of the mines, he began retamping the gallery. Gradually the bags of spoil were passed forward and quietly placed into position. As Woodward looked on, he could barely hide his apprehension. The addition of each bag made any repair harder. Fearing the Germans would find the mine and disarm it, from that moment on he had an officer working the newly installed switchboard, continually taking both continuity and resistance tests and reporting to him.

He had just returned to his dugout at 3 pm when the British artillery all along the front opened up with what was termed a 'demonstration barrage'. It frightened the life out of him, and he feared that the rumble and percussion might set off his mines. The roar of the shells continued for half an hour. The next day was noticeably quiet, but the following day at 3 pm the British heavy artillery gave the Germans another demonstration of their firepower, this time for an hour.

The time of the attack – zero hour – had been kept secret even from Woodward. It was not until late on 5 June that Brigadier General T. S. Lambert, the CO of the 69th Brigade, the British infantry who would attack the Germans at Hill 60, informed Woodward that zero hour was at 3.10 am and that he would be with him in the firing dugout to provide the countdown and to see the mines fired. To be told the exact timing was something of a relief, because at last Woodward could calculate what the expected location of the German tunnellers would be at zero hour.

Dawn broke peacefully on the morning of 6 June, the day before the attack was to start. Woodward recalled in his diary that it 'was a day during which one's nerves seemed to be strained to the breaking point'.[4] He had determined to fire his two mines with a Service Exploder and again checked the leads that came into his dugout from each of the two mines. He then carefully connected the two mines together into series and extended the leads back to the specially constructed, light-proof firing position in Bentham Road support trench, well behind the frontline. Now his leads were vulnerable to German artillery, stretching as they were 350 metres from the Hill 60 mine and 450 metres from the Caterpillar mine back to the firing point. It was something new to worry about.

It was mid-summer and the sun hung in the sky, sinking slowly and bathing the battlefield in an eerie glow. Woodward rarely saw the sun, let alone actually watched its path towards the far horizon, but as dusk came he stared at it, willing it to sink faster and just go away. By midnight, his work was complete, but the testing continued.

An atmosphere of peace and quietness settled over the battle-field. The Allied guns had fallen silent in the hope of not stirring up the enemy artillery. Were an artillery duel to start now, the infantry lying on their start lines along nine kilometres of front and packed into forward trenches would be exposed, resulting in horrendous casualties. The Germans also remained quiet, their artillery silent, their crews asleep beside their guns in the warm summer air. Overhead, the odd heavy shell fired kilometres behind the line travelled high across the heavens, on its way to the German rear.

At 2 am, a German flare arced up into the sky above Hill 60 and flooded the area with a dazzling white light. It hung there, then descended, leaving a smoky vapour trail behind as it fell to earth. Woodward glanced at his watch and gave the order for the troops in the mine system and guarding the tunnellers' dugouts to be withdrawn. They moved silently to their positions, well aware that on such a quiet night any noise would travel far.

At 2.25 am, with just 45 minutes to zero hour, Woodward made the last resistance test and confirmed to himself and to his two fellow Australians – Lieutenants Royle and Bowry – that their leads were still intact. All that remained to do was to connect the leads to the firing switch.

Just before 3 am, Brigadier General Lambert arrived and took up his position in the firing dugout. Lambert was one of the younger generals, 'A first class soldier who left nothing to chance and never hurried prematurely' and who was 'Close up to the scene of the action, from whence he would watch an attack and control by timely action its course'.[5]

The strain on the firing party was intense. Their great fear was that there would be nothing – no bang, no explosion, nothing – when the firing handle was depressed. All along the line, nervous officers were getting ready, their sweaty hands feeling for the cold brass of the plunger, praying it would all work.

Thousands of men were now watching the ticking second hand on watches along the front. Infantry officers, their whistles in hand, stood on their fire steps, a mass of helmeted men stretching as far as the eye could see, their bayonets glistening and their rifles ready. Thousands of artillerymen stood, their guns loaded, the firing lanyards in their hands, just waiting for the word to FIRE and send their first salvo. Crews sat sweating in their tanks, their hands on greasy levers ready to start up their lumbering giants. And everyone wondered about the awful day ahead and their likely fate.

In Woodward's firing position, all eyes were on Lambert. He stood, legs astride, the stopwatch in his open palm. No one moved. All eyes were on the watch, its black second hand continuing its relentless sweep, its ticking seemingly amplified in the silence. Woodward's hand closed silently around the plunger handle.

Then the silence was broken. It was Lambert. 'Five minutes to go.' One final check and the leads were fine. Woodward turned to Royle and Bowry. Each stood by an exploder at their feet, ready to spring into action should Woodward's dynamo fail. They were good men and he knew he could trust them.

'Three minutes to go,' called Lambert, his tone an octave higher than usual. Tick . . . tick . . . tick . . . tick . . . tick. All along the front, so much ticking, thought Woodward, so many men staring at slowly advancing second hands, so many Germans with seconds to live. 'Two minutes to go,' whispered Lambert. Woodward felt his hand tighten on the plunger handle.

'One minute to go.' Woodward took a deep breath as if it were his last. Then, 'Forty-five seconds to go', 'Twenty seconds to go' . . . tick . . . tick . . . tick . . . ten seconds, nine, eight, seven, six, five, four, three, two, one, FIRE.

Over went the firing handle. Woodward had grabbed it so tight and pressed it in so determined a way that his hand came in contact with the terminals and a strong electric shock sent him spinning backwards off his feet.

For what seemed an age, nothing happened. At first, far-off tremors rumbled through the earth from other mines fired along the line. In an instant, there flashed across Woodward's mind a feeling of envy for those officers whose mines had successfully fired.

And then a dull roar came from deep within the earth, growling and heaving, and suddenly the German frontline burst upward as a sheet of dark, muddy clay, planking, huge lumps of dirt the size of hay bales, and the cartwheeling bodies of men and weapons, shot skyward. Out of the top burst a sheet of red flame, highlighting the now cascading debris as it fell back and crashed and splintered in a wide radius.

As Lieutenant G. A. Hamilton, an artillery officer near Zillebeke noted: 'At exactly 3.10 am Armageddon began. Never could I have imagined such a sight. First there was a double shock that shook the earth here 5000 yards [c. 4600 metres] away like a giant earthquake. I was nearly flung off my feet. Then an immense wall of fire that seemed to go half way up to heaven. The whole country was lit up with a red light like a photographic darkroom. At the same moment, all the guns spoke and the battle began on this part of the line. The noise surpasses even the Somme; it is terrific, magnificent, over-whelming.'[6]

To the south, in the midst of the Messines–Wytschaete Ridge, the British Inspector of Mines, General Harvey, wrote in his diary: '3.10 am. A violent earth tremor, then a gorgeous sheet of flame from Spanbroekmolen, and at the same time every gun opened fire. At short intervals of seconds, mines continued to explode; a period which elapsed between the first and the last mine, about 30 seconds.' The 30-second lag, which had caused Woodward such envy, was due to the synchronisation of the officers' watches being slightly off.[7]

'With a fearful shuddering, the earth was shaken to the core,' wrote Oberstleutnant Otto Füsslein, the commander of the German miners along the Messines front. He saw vast craters open up, 'like the jaws of a hell-hound, spitting fire, water, earth and dark clouds into the sky. Clods of earth showered down over a wide area . . . and once more a hellish rain of steel poured out of the sky as though intent on annihilating every living thing below.'[8]

All along the line, 100,000 Allied infantry felt the earth heave and stared skyward as huge funnels of black earth climbed into the sky. Rising hundreds of metres, they suddenly broke open and from within, the same blinding flash of red flame scorched the heavens. And then the infantry's whistles blew, hardly audible over the crash of 2000 Allied guns along the whole of the British front, which sent an avalanche of shells into the shattered German line. The lighter artillery fired over open sights at the dark line of the ridge, now clearly visible and rim-lit by the breaking dawn. The heavier guns fell on predetermined targets: German artillery emplacements, which had been carefully and precisely located on maps, the lines of communication and advance for reinforcements, storage areas and billets, cook houses, rail lines and command centres.

About 460 tonnes of explosives, mainly ammonal, had been detonated, in the biggest man-made explosion in history. It brought people running from their houses in panic in Lille 35 kilometres away, fearing an earthquake. It was said that it was heard far away in London, and even in Dublin, 700 kilometres from Messines. It was also the most remarkable and successful mining feat in history and one that would never again be attempted or surpassed.

For a fraction of a second, Oliver Woodward, deep within his dugout, failed to grasp what had just happened. His shock was soon replaced by 'joy in the knowledge that the Hill 60 mines had done their work', he later wrote. 'We had not failed in our duty.'[9]

Woodward, Royle and Bowry shook hands and congratulated each other on their successful firing. General Lambert beamed, his lined face never happier now that he knew his men were advancing across no-man's-land, which only moments before would have been a suicidal task.

Eighteen

After the Earthquake

Deep within their dugout, Woodward and his officers had not seen the sight of the explosions. They made a hurried inspection of their mining system and saw that the dugout showed little damage and the upper section of the mine was also intact. He ordered an officer and a few men to carry out an inspection deeper down, then went across to what had been the German lines just a few hours before, to inspect the damage and establish machine-gun positions on the rim of the crater in readiness for the expected German counterattack.

Woodward measured the Hill 60 crater and found it to be 65 metres across and about ten metres deep. The radius of the destruction – which he called the 'rupture to obliteration' – was more than 40 metres, meaning that everything within a circle 80 metres in diameter was destroyed. The Caterpillar crater had a diameter of nearly 90 metres, a depth of nearly 16 metres and a radius of destruction of 58 metres.

Wandering further, he found the ground littered with shredded timber, lengths of metal and corrugated iron so torn and twisted that it was now almost flat. Trench walls had been squeezed together by the blast. The German dead were still standing upright, the trench sides having compressed so fast that they did not even get the chance to move. Beside them stood their rifles, some still upright. Along what had been their parapet were shattered sandbags; the contents had been atomised and the bags shredded and blown flat.

The history of the German 204th Division, which was occupying Hill 60, states that it lost ten officers and 679 men in the explosion. It seemed that two German companies had been in the process of rotation, with one going out of the frontline and one coming in, so that two companies were taken out. Those who somehow lived were nervous wrecks, crying with fear and throwing up their hands in surrender.

From the firing of the 19 great mines at 3.10 am on 7 June, the offensive along nine kilometres of the front around Messines was seen as a great tactical victory. By nightfall, the British objectives had in most cases been won, a most unusual outcome. The Battle of Messines was the most successful offensive up until this time in the war.

Great sections of the German frontline were destroyed, especially strategic strongpoints and the small forward salients that had bulged into the Allies' front. The German defenders were at first paralysed. At German garrisons all along the ridge line, those who were not dead were stupefied, dazed, nervous wrecks. The attacking Allied infantry found the Germans totally demoralised, some crawling about, others staggering, many crying and offering no resistance, their hands held high in surrender as they mumbled '*Kamerad*'.

The Australian 3rd and 4th Divisions attacked at the very

southern end of the line. Some drove through the village of Messines, while others attacked through the mined area just north of Factory Farm and Trench 122 near Ploegsteert Wood. 'Everywhere after firing a few shots the Germans surrendered as the troops approached,' wrote Charles Bean. 'Men went along the trenches bombing the shelters, whose occupants then came out, some of them cringing like beaten animals.'[1] William Garrard, a lieutenant of the 40th Battalion, Australian Imperial Force, noted, 'They made many fruitless attempts to embrace us. I have never seen men so demoralised.'[2] Although the figures vary, it is believed that as many as 10,000 Germans were killed in the exploding of the mines, and thousands were captured and wounded.

When the British troops had secured the Messines Ridge line, officers from the tunnelling companies were quick to move in and inspect the German workings to determine the extent of their mining operations. In many cases there was little to inspect, for the entrances had caved-in or vanished.

In the German press the following day, the focus was not on the collapse of the German line and the extent of the defeat, but on the British losses and the fact that the German second line, known as the *Oosttaverne Line*, had stopped the advance and remained in German hands. And, in truth, Allied casualties were slightly higher than those of the Germans. Some men, including the Australians, were delayed by the slowness of the British advance and left in exposed positions on the Messines ridge. Later they were caught by Allied artillery and forced to retreat, again with many losses.

Nevertheless, the Allies had made a great strategic gain: the high ground was no longer in German hands, and this was to prove very significant to the future of the war, particularly along the Belgian front. The German Official History states that the

Messines offensive fully succeeded and that the salient 'had been lost with dreadful casualties'. The battle 'was one of the most dreadful and depressing experiences of the regiment in the war,' according to the historian of the 18th Bavarian Infantry Regiment, and General von Kuhl was to write of the battle that it was 'one of the worst tragedies in the world war'.[3]

Among the Allies there were sceptics and detractors of the mining operation, both before and after the offensive. In the days before the battle, some British commanders requested that their men go in without the assistance of the mines because the effects of the mine explosions could not be foreseen and they mistrusted the assurances they were given about the safety of their men. And then after the attack, a British division claimed that the mines were 'a definite hindrance in the attack'.[4] One of their brigades had not been warned of the mines, and the unexpected massive explosion had caused a great deal of alarm among the men.[5]

Another criticism was that more strategically important positions should have been mined and attacked. Many of the mines were started in late 1915, and over the course of 18 months there had been some changes to the frontline, but it was almost impossible to change the direction of a tunnel, even over a relatively short distance. Others felt that the mines did not inflict enough casualties on the enemy and that the benefits were outweighed by the amount of effort, money and Allied deaths spent in mining.

Some of the attacking troops reported that German mines went up behind them as they advanced, due to the German line projecting into the Allies' in certain places. This was especially the case near Ontario Farm, where troops advancing from the southwest saw a massive explosion to their left rear. Other troops complained that the craters left by the mines meant they had to split up and go around either side of the crater and that as they did, they bunched up and were easily shot down. There were very few instances of craters being defended by the Germans in the early hours of the attack, though, so the advancing troops were mostly spared being fired down upon from the crater rims.

General Ludendorff, the joint supreme commander of the German forces along with Paul von Hindenburg, had no doubts about the success and significance of the mining operation. He later wrote: 'We should have succeeded in retaining the position but for the exceptionally powerful mines used by the British. The result was . . . that the enemy broke through on the 7th June. The moral effect of the explosions was simply staggering; at several points our troops fell back before the onslaught of the enemy infantry.'[6]

The failure of German mining commanders and geological experts to anticipate and eliminate the Allied mines triggered an enquiry and a search for scapegoats. Many times over the previous year, the German High Command had sought reassurance from Oberstleutnant Otto Füsslein that the British mining system was being monitored and that he had the situation in hand. Füsslein did not realise the depth and extent of the British mining effort. Since the beginning, the Germans had been confident in their work and felt they held the initiative. At a tactical conference in early 1917, Oberstleutnant Füsslein had boastfully declared that 'the mining situation was thoroughly cleared up and the tunnelling arrangements were superior to those of the enemy'. He proclaimed that he was 'satisfied that the British had been outwitted and outworked'.[7]

Captured Allied miners had let slip that deep mining was under way, but Füsslein had been fooled by the fact that aerial reconnaissance had failed to find the blue-grey spoil from deep mining because it had been so well hidden by the Allies. By this time, too, the British blockade of Germany was preventing crucial supplies from arriving, leaving the German tunnellers without necessary mining apparatus such as electric pumps and ventilation equipment, which set their countermining back.

The German tunnellers had struggled with underground water and running sand, and had thought the Allies would, like them, be able to dig only at a shallow level. Their geologists had believed this to be true, and their hydraulic engineering experts had reported that without proper drilling rigs, electricity, powerful pumps and decent ventilation, it was impossible to mine deeply along the Ypres–Messines salient.[8] In actual fact, though, although their position at the foot of the ridge made the Allies vulnerable to the Germans actively draining their mines down the hill and into their tunnels and trenches, the Allies had less distance to dig to arrive at the blue clay, which was less difficult to tunnel through.

The morning after the attack, the geologists were assembled for a thorough dressing down by a German general who cursed them for their complacency and ignorance, and blamed them for the massive casualties his army had sustained. As a result, those over 40 years of age were sent back to Berlin in disgrace and those under 40 were sent to an active regiment in the frontline.

✕

It had been a long night and a long day for Oliver Woodward, his officers and men. They felt exhausted; the stress had drained their last reserves of strength. At 6 pm they were relieved, and with their few possessions – their weapons and, for the officers, the important exploders and synchronising watch – they set off for their base camp at Ypres. Woodward would never set foot on Hill 60 again. Back along the well-trodden track to Larch Wood they went, past the Casualty Clearing Station where many of the Australian tunnellers had sought medical attention over the last eight months, past the lines of graves at Transport Farm and on up the broken duckboard track to Shrapnel Corner. The men seemed light-headed and relaxed, as though the end of the week heralded a long weekend. They tramped along the railway line to Lille Gate and, upon entering the Railway Dugouts, received a tongue-lashing from the major there, who told them it was forbidden to enter Ypres this way for fear of drawing German artillery fire. Woodward wondered if the Germans did not have more to worry about than a party of three grubby officers and 40 filthy, muddy men returning to Ypres, way behind the lines.

That night, Oliver Woodward made a simple and poignant note in his diary:

> Thus, on 7th June, 1917, the dreaded Ypres Salient ceased to exist, and due to the combined effort of Infantry, Artillery, and Tunnellers, the Messines Ridge was brilliantly captured.[9]

The Messines Ridge may indeed have been brilliantly captured, but there was more drama to come for him and his men.

NINETEEN

A New Role
and a
New Danger

A little over a week after the detonation of the mines, Oliver
Woodward, his officers and 200 men took a wonderful break
from the war, travelling in bright sunshine in ten London double-
decker buses with the top deck open to the warm summer air.
Although they knew danger still lay ahead for them when they
returned to the front, for now they were on their way to the
British 2nd Army's rest camp for tunnellers, well behind the lines
in the village of Malhove on the outskirts of the French city of
St Omer.

The camp was in an open field between a railway embankment
and a wood. In the middle was a lagoon about 500 metres long
and 200 metres wide where the men could swim and row in a
couple of old punts they found. At night, they slept well, resting
in the knowledge they were now well beyond the range of the
enemy's longest guns. Here there was no military training and
their time was taken up with long walks, letter writing, gambling,
eating, drinking and trying some good local wines. Woodward
laughed when he saw Australian officers going through the time-
honoured process of elegantly sniffing the wine to better enjoy the

fragrance, then taking a sip before grabbing the glass and pouring it down their throat as if it were a cold beer at a bar back in Australia.

Far from the war, the rest camp gave Woodward and his mates a chance to explore rural France. He especially enjoyed late-afternoon walks through the countryside, past rows of white-washed cottages with thatched roofs; a wisp of grey smoke slowly rising from their brick chimneys. Then along the canals, and on past the occasional fisherman sitting in the dappled light of the towpath, while far off church bells rang out vespers somewhere in St Omer. On other occasions he visited the cathedral and was shown around by the ageing priest, or wandered through the lanes that criss-crossed the Clairmarais Wood. He missed Australia, observing in his diary that these woods were 'too dark and sombre to be really beautiful . . . and in this respect the open sunlit glades of our homeland are more appealing'.[1]

$$\times$$

Far to the north, other Australian tunnellers were fighting a desperate and very different war: this time at the beach. The front-lines stretched from Switzerland at one end to the North Sea at the other, a total of 640 kilometres. At the North Sea, the trenches went virtually to the water's edge, certainly through the sandhills and down onto the beach. The Allied line was 600 metres on the northern side of the Yser River in an area where the beach and sandhills extended nearly two kilometres inland before they struck marshlands and low flooded areas to the east.

When the British took over from the French in June–July 1917 they decided to mount an attack against German strong-points in the sandhills. The 2nd Australian Tunnelling Company, specialising in soft soil, was brought in and ordered by General

Harvey to dig a tunnel under the German frontline to a machine-gun position. Their work attracted heavy German shelling and an infantry assault, which drove the British and Australians back to the River Yser. German artillery smashed the three temporary bridges that provided access to the British troops north of the river and quickly pushed the British line back.

Meanwhile the tunnellers were fighting their own little war. A dozen were caught underground. They held out for 24 hours but surrendered when their tunnels were polluted by German smoke bombs and the air became unbreathable. Others who weren't caught took to the river and swam to the southern bank. Many of the British troops could not swim and were trapped on the northern bank, so two Australian sappers, Burke and Coade, grabbed a long rope and raced to help. Coade held one end, and Burke swam back across the river, secured the rope on the northern bank and helped the non-swimmers across before swimming back himself.

Another Australian tunneller, sapper James O'Connell, had been burnt by a flamethrower during the fighting, and as he lay among the badly wounded he realised he had to get back across the river or be captured. He climbed across part of the broken bridge and swam the rest of the way, but when he climbed out on the other side, he heard an English soldier in the river calling for help. He immediately dived back in and brought the man to shore, before fainting through loss of blood. For his action he was awarded the Distinguished Conduct Medal.

✕

As soon as the Allied front had been stabilised, GHQ reappraised the work of the tunnelling companies and their future part in the battle ahead. Tunnelling had been viable near Messines because the frontline had been stagnant for such a long period. Now, as the offensive moved against Passchendaele, there was no time for tunnels and mines, so the expertise of these fine men must be put to another use.

On 2 July 1917, Woodward and the tunnellers left their restful surroundings at Malhove, 'fully refreshed in both body and mind', and returned to the desolation of the frontline.[2] They tumbled from their London buses at their new camp in a field outside Dranoutre, eight kilometres to the west of Messines. Little if any mining work would be carried out along the front of the 2nd Army, to which they were attached. Instead, they were put to work building and maintaining roads. Across the shattered landscape they built corduroy roads of timber planks that stretched in long, straight lines. These were easily seen by German observation balloons and continually targeted by German artillery, making the men 'shell shy' as they traded one fearful occupation for another.

Woodward's unit was then sent south to construct dugouts in the craters at Hollandscheschuur formed from mines blown in the 7 June attack. Advanced Brigade Headquarters and a dressing station were being built into their sides. Next, they were posted to the Hooge crater, just north of the Menin Road, in preparation for the Allies' advance towards the village of Passchendaele.

Camped and building a road alongside the Menin Road, they were in for a lively and dangerous time. Close by was Birr Crossroads and a few hundred metres back towards Ypres was the infamous Hellfire Corner. Both were shelled at all hours of the day or night by the Germans. The roadside was littered with upturned wagons, dead horses, mules, and the remains of men and equipment. According to Woodward his section suffered 'fairly heavy' casualties. One night, eight of his men on their way back to billets after finishing a shift were killed and four were wounded when a shell hit their stationary truck between Birr Crossroads and Hellfire Corner. Here the men had comfortable billets, but the presence of a large Allied naval gun nearby meant they drew fire nightly as the Germans sent over bombers to destroy it.

The battle for the Messines Ridge had been a resounding success, but the real prize was the high ground that arced around Ypres and was dominated by the Passchendaele ridge to the east. To take this, General Plumer planned a series of small, limited offensives, known as 'bite and hold' operations that would result in four battles, all of which would involve the Australians. It began on 20 September 1917 at 5.40 am when a massive artillery bombardment started the offensive that was planned to take the Gheluvelt ridge and maybe, just maybe, allow for the long-awaited breakthrough.

With shells landing to the east of Westhoek Ridge, the Australians of the 1st and 2nd Divisions attacked through Glen-corse Wood, closely following their barrage and pushing the Germans back. Attached to the Australian infantry were three sections of the 1st Australian Tunnelling Company. Their job was to rebuild German pillboxes captured in the advance, cutting an opening in the front, sealing up the German rear entry, cleaning away debris and preparing it for Allied use. Woodward and his section were held back, much to their annoyance, to enlarge the Cambridge Road dugouts, quite close to Birr Crossroads.

The Menin Road offensive was a great tactical success. By the end of the first day, the Australians had taken their objectives and were at the western edge of Polygon Wood. Some Australians had even entered Polygon Wood, past their final objective and a place that would itself become infamous in the following week with the attack of the 5th Division AIF. As Charles Bean was later to record, 'The advancing barrage won the ground: the infantry merely occupied it, pouncing on any points at which resistance survived.'[3]

As an engineer, Woodward understood the landscape, the water table and, most importantly, the well-established drainage systems and graded watercourses that for centuries had allowed this waterlogged land to be productive farmland. However, the drainage systems were quickly smashed by the Allied bombardment. Now the water had nowhere to go and ran into the shell craters and trenches, and turned the landscape into a quagmire. Passage through this muddy wasteland was virtually impossible. The conditions were unbelievably atrocious, with men, guns, horses and tanks sinking into the mud and disappearing – due to the ignorance, obstinacy and blind stupidity of the British commanders, who had not grasped the terrain or the weather.

While the Australian 5th Division threw themselves into a battle, their first since being decimated at Fromelles over a year before, Woodward and his men had their work cut out keeping the roads open for the reinforcements, the food and medical supplies and the ammunition needed at the frontline. With the battle raging, the lines of communication were never free from strafing and relentless attack, the men never allowed to get organised, the roads never allowed to be repaired and no one allowed to feel safe. Two officers and 130 men worked on the road that ran over the water-filled craters from Westhoek Ridge up to Zonnebeke, alternating between one day's work on the road and one day's work bringing up road materials, timber and supplies.

In his diary, Oliver Woodward relates a story of an Australian supply column – one man riding a mule and leading another – taking ammunition to Anzac Ridge. When they were slowly meandering back down the narrow, muddy duckboard track, returning to the Menin Road, there was a massive explosion. One mule received a direct hit from a shell, and vanished. The man leading the mule was unhurt; he was more surprised about the sudden disappearance of the animal. Not swallowed up by the mud that engulfed so many corpses of men and beasts here, this mule had been blown to smithereens by the intensity of the explosion. Woodward was amazed. 'As a matter of genuine interest I examined the area and can truthfully say there was no sign of the horse, either in the form of blood or fragments,' he wrote.[4] To him it offered an explanation for the fate of men who had similarly been lost without a trace on the battlefields.

By the end of October, Polygon Wood was in Australian hands, but the casualties were high, with 5452 killed or wounded in the two Australian divisions. While the battle raged towards the village of Passchendaele and the infantry died in their thousands in the mud and slime, the tunnellers were ordered to construct winter quarters for the Allied armies along the Ypres front.

Into 1918 the work of the tunnellers continued, not as they were recruited and trained before the Hill 60 attack, but now as sappers and military engineers. In early 1918 Aboriginal tunneller Herbert Murray joined the 2nd Australian Tunnelling Company. He was one of approximately 400 Aboriginals who fought in the war, but there were only four Aboriginal tunnellers (the others – brothers Alfred and John Ponton, and Alfred O'Neill – were in the 3rd Australian Tunnelling Company).

Wherever there was something to be built or something to knock down the tunnellers were called upon. In January 1918 the local constable at Zonnebeke told the Allies that there was a crypt under the church that might make a good Allied dugout. Captain Woodward had his men dig a passage through the foundations and then a longitudinal drive – a short tunnel – across and under the church, but no crypt was found. However, all this digging did in fact make an excellent dugout, which served as a battalion headquarters and Casualty Clearing Station.

Much of the Australians' work was far more dangerous. On 13 March 1918, a shift of Woodward's men, moving up the Potijze–Zonnebeke Road came under a fierce mustard-gas attack. Fifty men of the company were severely affected and had to be evacuated. Mustard gas had been first used by the Germans in July 1917 near Ypres, and although the standard-issue British gas masks were quite effective against inhalation, mustard gas burned exposed skin, which blistered and became infected.

$$\times$$

On 21 March 1918, Germany suddenly launched *Operation Michael*, a massive offensive along a 100-kilometre front from Arras to St Quentin. Utilising the added firepower of an additional 63 divisions, withdrawn from the Eastern Front after the collapse of the Russian armies and the signing with Russia of the Treaty of Brest-Litovsk in early March 1918, the Germans planned to drive the Allies back and take Paris before the might of America could be brought to bear.

Following a five-hour bombardment by several thousand guns, the German infantry pushed the British back, forcing a demoralising retreat over the hard-won battlefields of the previous two years. The Germans hoped to drive a wedge between the British and the French fronts that met along the old Roman road that ran from Amiens to Peronne, take the large provincial city of Amiens, and then catch the train to Paris. Even if they failed, they could force upon Britain more generous ceasefire or armistice conditions and do this before the full weight of the American contribution really affected the balance of manpower and the outcome of the war.

Within days, the Australians were racing to various points along the front to stop the German advance. The 3rd and 4th Australian Divisions moved south. Near Hébuterne south of Arras, men of the 4th Division met French refugees fleeing westward away from the fighting. It is documented that upon seeing the Australians and recognising their strange hats, the refugees stopped their carts and began unloading. When asked why they were not running, the reply came back: *'Pas necessaire maintenant – vous les tiendrez!'* (Not necessary now – you'll hold them!)[5]

On 28 March, one week after the opening of the German offensive, Woodward, his officers and men were rushed south from their camp to the French village of Saulty, just to the north of the Arras–Doullens Road, 40 kilometres to the north of Amiens. Here they were to work on the GHQ Defence Line, a defensive line for the retreating troops to fall back onto and hopefully make a stand. Immediately they began work building trenches, machine-gun emplacements and mortar positions protected by barbed-wire entanglements.

One day, while hard at work, Oliver Woodward was confronted by the British engineer-general in charge of the construction of the defensive line. He demanded that the Australians work from dawn to dusk, but they were not going to have any of this and were quick to let him know. The British general worked himself into a frenzy and appeared to be on the point of bursting with rage. 'We expected to see an explosion compared with which the Hill 60 Mines would be insignificant!' joked Woodward.[6] He calmed the enraged general down by suggesting a way of working that would increase the men's output by 50 per cent and the officer was satisfied. The Australians got to work and made good progress. 'Later when watching our men at work he was man enough to express surprise and unstinted praise for our efforts,' wrote Woodward.[7]

The tunnellers got stuck into their trench digging, their wiring and their defensive work, well aware they were turning the lovely French countryside into another battlefield. While Woodward was pleased with the 'fields of fire' he had created for the machine guns, he was also disturbed that the beautiful woods of Chatalet and St Pierre were suffering the same fate as the shattered woods he had seen in northern France and along the Menin Road in Belgium.[8]

The men worked hard, but they still managed to uphold the Australian soldiers' reputation for skylarking. They were billeted at a nearby farm, and one day a man was running along the narrow wall of the farm's waste pit when he tumbled in, landing fair and square in the slimy green water. Woodward and the officers, sitting quietly in the farmhouse with the windows open, were suddenly overcome by a terrible stench. They rushed outside to find the poor sapper pulling himself from the pit, covered from head to toe with the most vile muck, to the mirth of the men and the officers alike. Woodward dryly remarked, 'This episode cost the Australian Nation a complete Uniform, and I think it was many a day before that Sapper got rid of the smell.'[9]

Early one morning as the men were about to commence work digging a trench, they saw an old French farmer planting potatoes by hand. The ground was wet after a period of rain, and a low, wispy fog hung over the ground. Wishing to augment their awful rations with some potatoes, and the temptation being too great to resist, the men quietly followed the farmer, digging up the potatoes as fast as he planted them, collecting in the process about

70 kilograms. When the news of this stunt leaked out, the farmer demanded compensation, which was covered by the company's canteen fund.

In early June, Woodward and his men were sent further north, to Bouvigny-Boyeffles, very close to the large regional city of Lens. Now for the first time in many months they were back under the German guns. At night they could look from the ridge behind the town and see the frontline illuminated with flares, and behind the lines, the sheet lightning of the artillery as the nightly duels were played out. 'It looked like a beautiful pyrotechnic display,' wrote Woodward, 'that is, to those who could view it from a safe distance. To the men in the line, it was a very different story.'[10]

They were given the job of excavating a set of stairs from a trench down into an ancient underground chalk pit. Over the centuries, blocks of hard chalk, used in building materials, had been cut out and removed, while pillars had been left to support the roof. Now there was a maze of caverns, which the Allies were planning to use as underground shelter for their troops.

By the end of June 1918, the German advance had been halted and the lines re-established. The Allies had begun planning for the push back towards Germany, now supported by American troops in large numbers. In the three months they were in this area, Woodward's section of 110 men had worked on a frontline of 34 kilometres, had built four kilometres of GHQ Defence Line and erected 30 machine-gun emplacements. They were proud of their achievement.

Woodward suffered a bout of malaria on 30 June and was sick for four days. 'On the 5th July my leave pass came through and I staged a remarkable recovery,' he dryly noted in his diary.[11] When he returned, the next task for him and his men was constructing artillery positions between Bussy and the village of Querrieu just to the east of Amiens, where they arrived on 30 June 1918.

Then, with a major Allied offensive about to start, they built a large Brigade Headquarters dugout between Hamelet and Le Hamel, very close to where the Australian National Memorial now stands. They were also detailed to remove time-delay mines and search for booby traps left by the Germans. On 8 August, the great attack started, spearheaded by Australian, Canadian and New Zealand troops who pushed east along the Amiens to Peronne Road in what became known to the Allies as the Battle of Amiens. For the Germans it would be referred to as 'the black day of the German Army'. By the end of the battle, four days later, the AIF were digging in on a new line near Proyart. In some places, the frontline had moved forward ten kilometres and the Australians had captured 8000 prisoners, 350 machine guns, 40 trench mortars and over 80 field guns. The following day, the new commander of the Australian Corps, General Sir John Monash, was knighted by King George V, the first time in 200 years that a British monarch had knighted a commander on the field of battle.

'We felt that once again we were on top, and that we would eventually win the War,' Woodward wrote. 'How and when this end would be reached did not concern the individual. The essential fact was that one and all recognised we had the game well in hand and eventually the winning goal would be kicked. Our whole view-point had been changed in a few hours.' The following morning, he inspected the area that had been captured and found it hard to imagine, 'except for the sound of the guns ahead', that just 24 hours before this had been German territory and far behind their lines.[12]

For Woodward and the Australians the success of the attack

relieved months of tension. After the successful retaking of Villers-Bretonneux in April–May and the attack at Le Hamel in early July, there was a residual fear that the Germans would launch another massive offensive and that it might be more successful than Operation Michael. They could not foresee that offensives by France in July and an American attack against the enemy's vulnerable salient in Champagne would prevent such a move. Nor could they have known that by late July, it was clear to the German generals that there was now no hope of winning the war and the best they could do was establish a line and work for an attractive ceasefire and armistice.

From early August until late September, Woodward's tunnellers worked to support the rapidly advancing Allied front. They worked with the Canadians to repair the railway yards at Villers-Bretonneux, then moved eastward, building roads, repairing water supplies, dismantling German demolition charges, fixing railway lines and erecting pre-fabricated metal Nissen huts and accommodation.

While camped outside Cartigny, Woodward and his men were attacked nightly by German bombers. The British searchlights and anti-aircraft guns proved totally ineffective. Night after night the Germans came, bombed and strafed the Allied lines and flew off without any problems. One particular night, however, the tables were turned. When the enemy planes came over, they were picked up by the searchlights as usual, but this time British fighters were circling at a high altitude. They swooped down on the German fighters caught in the beam of light and, with their machine guns rattling, quickly shot down two of the enemy, who went spinning and burning to earth.[13]

✕

The Australians now increasingly found themselves working alongside American troops. Woodward wrote in his diary of being at a temporary camp in Tincourt Wood when a group of Americans – five officers and 20 men – arrived, without the faintest idea where they were or where the rest of their regiment was camped. Woodward fed them and helped them on their way, but it was the first of a number of occasions that would make him contemptuous of their training, tactics and competence as fighting troops.[14]

On 26 September 1918 Woodward's section had completed work on a road from Tincourt to St Emilie, in preparation for an American attack on the Hindenburg Line planned for 27 September. An American engineering company was assigned to assist him, and together they completed the roadworks. The attack was a failure as the American troops did not 'mop up' after themselves by eliminating any Germans who had been missed when the attack had gone through. As a result, the remaining Germans simply waited in their deep bunkers until the Americans had passed, then re-emerged, set up their machine guns and fired into their backs. Woodward was appalled:

It was tragic to observe the state of disorganisation which existed among the American troops. Probably as individuals they were not to be blamed, but their behaviour under fire showed clearly that in modern warfare it was of little avail to launch an attack with men untrained in war even though the bravery of the individual men may not be questioned. In effect so far as the Americans were concerned it was a case of a mob let loose, all plans forgotten and no definite objective in view. No wonder the German Machine Gunners had a field day. They must have felt like poor sportsmen shooting sitting game.[15]

Woodward was further angered when he later discovered that the Americans had erected a memorial to the men of the Tennessee Brigade 'who broke the Hindenburg Line'. Had it been a British or French memorial it would have been erected in memory of the men killed, rather than making 'absurd claims as to the part they had played in winning the War. If the American Nation had any self respect, it should demolish this monument to their bragging, and not ridicule the men of Tennessee who gave their lives to the common cause.'[16]

When the Australians attacked on the 29th, their artillery could not provide a creeping barrage, for fear of hitting the Americans, who were now sandwiched between the main German army and a line of machine gunners they had missed in their advance. When the attack commenced, the Australians, including Woodward's own men, 'suffered terrific casualties – Australians murdered simply because the Yanks failed to obey orders'.[17] Just after the attack started at 5.50 am, Woodward, along with 82 Australians and 65 Americans, began road repair work close to the firing line, at Bony. As they approached Quennet Copse and Guennemont Farm, they came under machine-gun fire. One officer and a sapper were killed, and 20 others were wounded. Receiving orders to inspect the condition of the road towards Hargicourt, he went out with a sapper to inspect it, in the midst of artillery and machine-gun fire.

Woodward had shown great bravery and leadership that day, and he was awarded a Bar to his Military Cross. 'By his courage and resourcefulness in patrolling the roads and organising the work he succeeded in carrying the work forward, thus enabling the subsequent attacks to be carried through. He set a fine example to his men at a time when casualties were heavy,' read the citation.

The Allied advance was now pushing forward, in some places so fast that only the cavalry could keep up with the retreating Germans. Moving with the advancing troops, between La Haie Menneresse and St Souplet, Woodward came upon a long line of Americans killed by Allied artillery fire. The plan had been to withdraw the infantry back 500 metres and then the Australians would lay down a creeping artillery barrage. 'The American staff failed to carry out the plans, with the result that our own barrage dropped right on the Americans, and they were killed in hundreds,' wrote Woodward. 'The bodies lay in a rough line marking what was the original Front Line position.'[18]

After their harrowing time on the frontline, Woodward and his men had ten days' rest in Cartigny, well behind the line. They returned to Becquigny, and on the night of 29 October their bivouac was bombed by German aircraft. A bomb landed right in the middle of the tent lines, and five men were killed and a further eight wounded.

TWENTY

The Last Battle

It was just ten days before the Armistice and the end of the long, dangerous war. But as is often the case in war, there was one more job to do and one more chance to get killed.

On 1 November 1918, Woodward's section was attached to the Royal Engineers of the British 1st Division. Their task was to build a bridge at a lock on the Canal de la Sambre à l'Oise in three days' time. And this was no ordinary bridge. It had to be strong enough to support a 34-tonne Mark V tank . . . and it had to be constructed under enemy fire.

The original bridge, which had crossed the canal 50 metres from the lock, had been destroyed by the retreating Germans. Now the only place to cross was at the narrow lock wall, whose gates were jammed slightly open. They had fortified and set up two machine guns in the boiler house near the lock gates and were also manning the adjacent pump house and the lock-keeper's house. The plan was for the British infantry – the 2nd Battalion, King's Royal Rifles on the right and the 2nd Royal Sussex Regiment on the left – to storm across at the partially opened lock gates, take the fortified boiler house and eliminate any resistance.

They were then to form a perimeter on the German side of the canal to allow the bridging detail to do their work.

The Australian Infantry Divisions had long gone from the front. Germany and its allies had been collapsing for a couple of months. The papers were full of the news that they were seeking peace treaties. The German Navy in Kiel had mutinied, and some elements of the army were in open revolt. In Berlin there were food riots, widespread acts of civil disobedience and people openly burnt pictures of the Kaiser. With the end of the war so close, it was not a time to risk one's life building a bridge.

The task of leading the construction of the bridge had fallen to Oliver Woodward because, 18 months before, he had attended a bridge-building course at Aire, a canal port to the west of Lille. It had been realised that as the Allied armies pushed towards Germany, the enemy would destroy the bridges as they retreated, so bridge building became an important function of the engineers. The bridges needed to be pre-fabricated, and simple and quick to assemble, but they also had to be able to take heavy loads, especially the tanks that were now an important part of the Allied armies.[1] Towards the end of the course, he was given a practical test of rebuilding Bridge No. 2 over the Haute Deûle Canal. He, along with another officer, had three days to complete the plans and schedule to the most minute detail to undertake this work, something that prepared him well for the task ahead.

Woodward blamed his bad luck at being given this job on his knowledge of bridge building. But he blamed the need to erect the bridge under fire on the Americans. According to his diary, the Americans had failed to press home an attack there. They had not driven the Germans from the bank, and this had allowed them to dig in and fortify their side of the canal. As a result, crossing had become dangerous for the Allies. Though proud to be working with such a famous and celebrated British division, Woodward knew he and the men were in for a perilous time.

Woodward was briefed by the Royal Engineers on the afternoon of 2 November. At 6 pm, as rain poured down, he visited the Tank Corps depot at St Benin to measure the width of the tanks and their tracks and to check on their weight to make sure the bridge could support it. From here he went to check that the necessary bridging material was available, and it was not until 10.30 pm that he returned to camp exhausted.

He woke the next day in the dark, shivering, his feet frozen, the flimsy, flapping tent offering no warmth. The weather was cold and overcast, the rain close to freezing, the ground muddy and soft. And the next two days offered grim prospects and dangerous possibilities. He left camp at 7 am and inspected the work of his carpenters, who were preparing the bridge's decking. Satisfied with this, he moved on to examine the steel fabrication and the welding that had been done overnight.

Woodward had heard of cases where a man had a premonition of his own death before a battle, asked a mate to send his last letter and personal effects home and then lost his life in the engagement. And as the men assembled to leave camp for the canal, he was approached by Corporal Albert Davey, who asked if he would post his personal belongings to his wife at home should anything happen to him. Woodward considered Davey, a 33-year-old married miner from Ballarat West who had been with the company since early 1917, one of his best non-commissioned officers. Davey said he'd had a premonition that this was to be his last action, that he would be killed.

Woodward was himself concerned about what might lie ahead. 'I frankly admit I was more disturbed by this turn of the tide than at any stage in my War Service,' he wrote. 'The query which kept coming before one's mind was, "Shall I be fated to come through safely?" Yet he tried to lift Corporal Davey's spirits, assuring him he would be fine. The corporal persisted, saying to Woodward, 'Captain, nothing you can say will remove the conviction that I will be killed. Will you please do me the favour I ask?'[2] Woodward replied that he would and took the man's possessions for safe-keeping.

Woodward went over the construction details with the CO of the Royal Engineers, and after a final briefing the tunnellers and the infantry set out for the lock. They headed north and reached Mazinghein just on dusk. Leaving their motorised transport, the men marched southeast to the small village of Rejet-de-Beaulieu. The road was swept with enemy artillery fire and a number of English infantry were wounded. Before Rejet-de-Beaulieu, Wood-ward took a side road, skirted the village and led his men to a point about a kilometre from the lock, arriving at 8 pm.

On schedule, the steelwork pre-fabricated in the Engineers' workshops arrived soon after by heavy horse-drawn wagons. They too had come through the shelling of the road and, with the horses skittish and edgy, the drivers were keen to unload and get out of the firing line. An enormous and daunting pile gradually rose: five tonnes of steelwork and girders, six metres in length. Also arriving on the wagons was the sawn timber that made up the bridge decking.

A cold drizzle set in. The rain was the last thing the men needed. Their hands were numb and the water trickled down their backs and filled their boots as they contemplated their task: to move this pile of steel and timber to a staging post 300 metres from the German fortified boiler house. The night had come down pitch black as they laid their cold fingers upon the freezing steel. One efficient lift and the first 360-kilogram steel girder was balanced on the shoulders of ten men. They stepped forward gingerly, feeling with their feet for rough ground. They knew that just before they arrived, the narrow dirt lane they had to walk along had been shelled; it was smashed and cratered.

Slowly they stepped forward into the unknown, slowly they tested the ground with their feet before planting their weight. Should one man fall or unbalance the others, the heavy iron girder could fall and crush them. Even if a shell burst close by, they must not lose their balance. On they went. A shell exploded 50 metres ahead and shards from the explosion whizzed over their heads and disappeared into the distance. Then a machine gun opened, a stream of tracer searching out attackers somewhere out in the darkness off to the right. As one team of men moved off, the next stepped forward, hoisted a girder and set off.

Slowly they edged forward. The road was dotted with potholes filled with water and loose, churned-up soil, creating a sticky, slippery mud. The weight of the girders bit into their shoulders, their collarbones ached and their fingers froze to the icy steel. A slow step, a re-balance, a glance up at the dim horizon, and then heads down as the cold rain beat into their faces. Somewhere ahead were the Royal Engineers who had moved off earlier to build a narrow footbridge across the lock. The Australians envied their task. It was more dangerous, because they had to build the footbridge under fire before the British infantry went in – but they were carrying only a light timber bridge, in handy, easily managed bundles.

Exhausted, the men finally made their assembly point, a sunken road that ran parallel to the canal. They put down their loads, rubbed their sore and aching shoulders, and prepared to lie down and wait until it was time to move forward to the lock. But Woodward was apprehensive. Sunken roads were known assembly points, making them favoured artillery targets. It was time to leave. Neither his officers nor his men had the strength or the inclination to move, but when he quietly explained their predicament, they dragged themselves to their feet. They respected his authority and experience.

Woodward led the men out into an open field a couple of hundred metres from the sunken road and selected a position for them to lie up until the Germans had been taken out by the British infantry. Quietly they dug themselves shallow trenches or crawled into shell craters – and watched as shellfire hit the sunken road where they had been assembled. The following day when they returned along the sunken road, they would find the ground strewn with the bodies of dead English troops who had been caught in the German bombardment just minutes after they had left.

As the rain came down, the men settled into their shallow trenches and muddy shell scrapes – small dish-shaped holes scraped into the earth, each big enough to give one man some shelter from exploding shells – and tried to sleep. As a thick mist came in from the canal, the men shivered. Their clothes saturated, some lay in the mud shaking, as a kilometre away across the canal a German field gun began a slow traverse. Its crew fumbled with a cold shell, slid it into the breech and pulled the firing cord. The shell arced across the canal and crashed to earth among Woodward's men. Earth and mud flew up, and the rank smell of cordite filled the misty air. A low moan came from the smoking earth as two wounded men rolled around in agony. Nearby lay the prone and lifeless body of Corporal Davey, his destiny fulfilled.

'This incident left me somewhat dazed . . . In temperament Corporal Davey was of the calm type, he was a soldier who was as fearless as any soldier can truly be in War, had never failed in carrying out his duties in a most efficient manner and inspired the confidence of his men,' wrote Woodward. He believed it was not fear or panic that caused Davey to approach him before they left camp and ask that his personal belongings be sent to his wife. 'He knew his call was coming.'[3]

At 4 am the German guns fell silent. The mist shifted, swirled and broke up, showing the tunnellers spread out among the shell holes and their shallow trenches. Woodward did a quick count of his men then called for stretcher-bearers to take the wounded to the temporary Casualty Clearing Station situated near the village.

At zero hour, 5.45 am, a British-artillery, machine-gun and Stokes-mortar barrage crashed onto the German side of the canal. The sound of incoming shells could be heard, fired from beyond Rejet-de-Beaulieu. Ahead of Woodward's prone men, four British 18-pounders fired directly at the three buildings occupied by the Germans across the lock. Soon, the German artillery opened fire. The shells were concentrated on the sunken road and the advancing British infantry, but again some landed in the open field among Woodward's men.

The British engineers edged towards the tree line just 30 metres from the lock and lay there waiting for their orders to storm the boiler house. Bent low, Woodward and Sergeant Hutchinson crept forward, and arrived just in time to see Major Findlay lead his men from the 409th Field Company across the lock. As the British barrage lifted, the men raced from the cover of the trees, over the exposed grassy verge and across the partially opened lock gates. The German machine gun located in the boiler house barked and bullets whistled off the concrete and steel, sparks flying. The men followed Findlay, who crossed the canal first, hurling bombs at the boiler house as he went. The German gun fell silent but the ground was strewn with English dead and wounded. For this action Major George Findlay, MC and bar, was awarded the Victoria Cross.

By 7.30 am, it was considered safe enough for the Australians to begin their bridging of the canal. Although the German machine guns had been silenced, intense artillery fire was landing on both sides of the canal and the narrow dirt road that materials still needed to be brought up. The Germans knew that the Allied bridging operation was under way. Through this bombardment came Lieutenants Sawyer and Thomson leading the tunnellers, who manhandled the long, heavy bridge sections up to the side of the lock.

Hutchinson and Woodward raced across the exposed grassy verge and leapt across the canal gates. They quickly prised off the coping stones inlaid along the top edge of the canal and squared off the rough angles. This would allow a snug fit when the first girder was dropped into place. As it was being eased across the gap, suddenly a small-calibre shell fell among the men, sending three spinning backwards wounded. Then the second girder was ready and this was quickly slid into place, a task now much easier with the first one in position. The Germans had lifted their bombardment so that their shells were falling 100 metres forward, where the British infantry were moving out into the open fields.

Across the two laid girders, the men could attach the cross members and then lay down the timber planking. In just two-and-a-half hours, the bridge was ready and the first tanks began crossing the lock. It was an amazing feat. Sadly, five Australians had been killed and another five wounded, just a week before the end of the war.

At 11 am, Major Findlay ordered the Australians to return to the village and rest. With them went the bodies of their dead mates, back along the muddy, narrow lane and back past the shattered infantry bodies in the sunken road. With them went the padre of the Royal Engineers who conducted a burial service for the Australians. Their bodies were laid to rest in the village cemetery at Rejet-de-Beaulieu. The tunnellers bowed their heads and said a last goodbye to five brave mates who had survived so long and contributed so much and were now in the cold earth, lost to their families back home.

The men, tired from a strenuous and dangerous night and morning, marched back to Mazinghein, where they rested until a convoy of trucks arrived to take them back to camp. Woodward reflected in his diary:

> As we lay in that grassy field on a bright sunny day, there was an atmosphere of sadness, more pronounced than usual. I feel that this was entirely due to the fixed belief that we had taken part in the last staged battle of the War, and this thought carried our minds to our comrades who, as it were, had just been given one glimpse of the long expected Armistice, only to lose their lives before it materialised. It was a matter of a few hours, but it was not to be.[4]

The men received accolades from their commanders. General E. P. Strickland, the commander of the First Division, in the Special Order of the Day praised their 'cool gallantry', and the commanding officer of the 1st Australian Tunnelling Company, Major E. S. Anderson, commended their 'very gallant conduct'. He went on to say: 'The Section, under the command of Captain Woodward, displayed such great courage, devotion to duty, and disregard to personal risk.'[5]

Woodward was praised for his 'complete disregard of personal danger' and 'great gallantry' by Colonel C. E. Sankey, the commanding officer of the Royal Engineers, First Division. Sir Douglas Haig's Dispatch of 8 November 1918 noted his 'gallant and distinguished service in the field'. And at a ceremony in Australia in 1920, HRH The Prince of Wales would award him a second bar to his Military Cross, for his 'conspicuous gallantry and devotion to duty'.[6] It was one of only four second bars awarded to Australians during the Great War.[7]

Lieutenants Thomson and Sawyer were each awarded a Military Cross for the operation, while Sergeant Hutchinson received the Distinguished Conduct Medal.

This was the last action fought by Oliver Woodward's tunnellers and the last action fought by Australians in the Great War. Just one week later, on 11 November, the guns fell silent, an Armistice was signed and the killing stopped.

Early on the morning of the 11th, a fine and clear day, the Tunnelling Company were called to parade. With deep emotion, his voice breaking, the commanding officer announced that at 11 am hostilities would cease. Germany, he said, had signed an unconditional surrender. The Germans were to withdraw within 14 days from all captured territory and within a further 16 days to ten kilometres beyond the Rhine into Germany. All Allied prisoners of war were to be returned, but German POWs in Allied hands would remain in captivity to prevent them re-forming a fighting force. Germany was also required to hand over the majority of its weapons, fighter aircraft and naval vessels, and transportation.

He thanked the men for their contribution, but added that there was still work to be done before they could all go home. Woodward recalled: 'The outward manifestation of joy which could be expected on such a memorable occasion was absent. We were as men who had completed a task which was abhorrent to us. The occasion called for thanksgiving.'[8] Around the world, people crowded the streets to cheer and celebrate, sang patriotic songs and gave thanks. Oliver Woodward and his section, like many men still on the front, reacted very differently. They had signed up for the period of the war plus four months. For them, the stage was still set for war. 'This probably accounts for the difference in reaction between men in the Front Line areas and those in the Cities and Towns of our Empire,' Woodward wrote. 'Instead, officers and men moved quietly about from one group to another giving and receiving a handshake among comrades. It was an occasion too great for words.'[9] Many quietly said a prayer of thanks for their delivery from this awful war, and their minds now turned to their families and their loved ones so far away back in Australia.

TWENTY-ONE

Going Home

Oliver Woodward's men had special skills, not the least being an understanding of explosives and a knowledge of German delayed mines and booby traps. So it was that over the next two months they travelled all over northeastern France and the eastern border regions of Belgium removing mines. They were attached to the 9th British Corps and proceeded with the advanced guard of the 1st Brigade of the 1st Division. It was seen as a great privilege to be attached to such an old and honourable regiment.

The Germans provided maps on which the positions of the mines were marked, and full particulars about how their mines worked. Most were fired conventionally with an electric detonator and were easily disposed of, because they were not likely to fire without an electrical current. A small percentage was set off by delayed-action fuses, and disposing of them was highly dangerous. To set these mines, the Germans would dig a two-metre shaft and into this place heavy-calibre shells. To one shell they would fit a special nose cap that held a small container of acid through which passed a wire. The acid would eat away the wire, and this would cause the release of a percussion spring that would fire the detonator and explode the mine. By varying the thickness of the wire, they could crudely time the length of the delay. To render the mine harmless the nose cap had to be removed, and this was dangerous work, as the mine could explode at any time. 'In carrying out this work, I honestly believe we were more nervous than any time in the war,' Woodward revealed.[1]

He thought it unfair that such dangerous work was put upon men who had already served so long in such dangerous conditions as the tunnellers. 'I was astounded to learn that at this stage of the War we were to undertake the task of removing delayed action mines . . . It was over the odds that we should be called upon to undertake such work, but as we were still on Active Service we had to carry out our duty.'[2] He believed that the French had the right idea: they made the Germans clear their own mines.

Travelling eastward, towards Germany, Woodward came to the French village of Thuillières, which had only been vacated by the Germans three days before. Here he removed mines laid in the town square. The relieved and grateful townspeople assembled and, with the band playing, cheered the Australians. The crowd was astonished to learn that the Australians had come from so far away and volunteered to fight the Germans in a war that had little to do with them.

That night, Woodward received 'one of the few Military Orders [he] was delighted to obey': from then on, they were not to remove delayed-action mines, but just to locate and mark them.[3]

As Woodward and the men headed towards Germany, marching along country roads, he passed many recently released prisoners of war slowly marching westward. These men were of many nationalities and many armies, their clothes in tatters, their boots worn out. In most cases, they were hungry and destitute.

'The contrast between these men and ourselves was simply astounding,' noted Woodward.

> We were a body of men who had experienced our fair share of the horrors of the War, and probably showed signs of our 2½ years' Active Service in the battle zone. Yet compared with these prisoners of war we were a crowd of fresh young school-boys. It made one realise to a small extent what sufferings these men had experienced . . . It was a tragic sight, and I felt more harrowing to look upon than the dead on a battle field.[4]

When the Armistice came, many were simply released, much like sheep, and allowed to wander off. There was little Woodward and his men could do, apart from sharing their rations or a little money and directing them to Beaumont, where help was waiting.

In late November, Woodward received orders that the Australians were to return to the Company Headquarters, at Marbaix, near the Belgian border south of Mons. No Australians would be part of the occupying forces that were to cross into Germany. This was a great disappointment to the men.

Then in late December, Woodward was ordered to take a small group of men and investigate a possible mine in the railway tunnel that ran from the Belgian town of Stavelot near Spa, to the fortress town of Malmedy, in the east of Belgium close to the German border. With seven men, including his trusty sergeant Hutchinson, he travelled to Stavelot on 20 December. The station master was very pleased to see him.

His men were provided rough accommodation in the station buildings, but Woodward found himself a comfortable old-world hotel where he was able to enjoy a nice roast dinner, a few sherries and the conversation of an English officer beside a fine log fire burning in an open hearth. It was snowing heavily that night. 'The comfort accentuated by the snowstorm raging outside, I felt there had never been a war,'[5] he wrote.

The next morning, Woodward and his men arrived at the station early, where they met the station master and some officials anxious to have this nightmarish bomb quickly disposed of. They went into the long tunnel and there, sure enough, was a bunch of leads projecting from a block of concrete. At least, thought Woodward, it was an electrically fired mine and with care could be quickly removed. He called for Sergeant Hutchinson, who very cautiously began to chip at the concrete block to expose the wire. Suddenly the block fell away from the wall, landing with a crash at their feet, frightening the already tense Belgian railway officials.

Woodward slowly bent down and with his torch examined the block and the electrical cable protruding from it. Then he realised it was just a piece of facing concrete from the lining of the tunnel. A length of cable had somehow been caught in the concrete when the lining was applied, and had been there ever since. Woodward broke into laughter, and then so did the tunnellers, but it was some time before the traumatised station master could see the funny side.

He did not know it then, but this was to be Oliver Woodward's last duty as a tunnelling officer. He later reflected in his diary that much of his engineering training at Casula camp had been farcical, so it seemed quite fitting that his final task should be equally ridiculous, beginning and ending his military career on a comic note.

As the return train did not leave Stavelot until early in the afternoon, Woodward decided to take his small party and fulfil their dream of crossing into Germany, which was three kilometres away. They marched along a narrow snow-covered road until they came upon two timber poles on either side of the road painted black and white, marking the German frontier. So that no member of the party of eight could claim he was the first man into Germany, Woodward lined them up and together, line abreast, the men slow marched, 'with a modified goose step', across the border and into Germany. They were probably the first of a small number of Australians, other than prisoners of war, to have entered Germany. 'After all the years of War there was a feeling of satisfaction in having set foot in the land of the Hun,' wrote Woodward.[6]

After spending a cold, snowbound Christmas in Belgium, Oliver Woodward was granted twelve days leave and on 19 January 1919 left Charleroi bound for London. Little did he know as he gave his officers and men a quick cheerio that he would not be back and it would be years before he saw some of these men again.

When he arrived in London, he called at the office of the Mount Morgan Gold Mining Company. Here he found that the Australian head office had cabled London requesting that every effort be made to have him repatriated as soon as possible, as he was required by the company for their mining operations in north Queensland. Armed with a letter from the Chairman of the Board of the company, he went to the Embarkation Branch of the AIF to secure a berth. But getting a berth and jumping on a boat to Australia was not as simple as it might sound. Quite simply, there were not enough ships for the task of repatriating the men of Australia, America, Canada, New Zealand, South Africa, India and other countries. To further complicate matters, the Australian government demanded that the men must only travel on ships of a high standard, with ample space, good amenities and decent catering.

In the end, the solution was a very Australian one: the spirit of fairness must prevail, and so the 'first to come, first to go' policy was introduced. The men who had served the longest were the first to receive their ticket home, but two other criteria were also taken into consideration: if a man had a family or a job to return to, he moved higher in the queue.

Woodward satisfied one of the three criteria for early repatriation: he had a job and a letter to prove it. This was reasonably rare. Most men had no job to return to, and many had no profession, trade or vocational training at all, having joined the army straight from school.

After weeks of fruitless meetings, revolving doors and waiting on answers, and after getting a further extension to his leave to allow him to stay in London, Oliver Woodward finally received a cable informing him he was booked on the transport *Czaritza*, which was to sail from Portsmouth on 17 March. He felt sorry that he had not declared his thanks and said his final goodbyes to the men back in Belgium before he had gone on his leave.

He boarded the *Czaritza* on 16 March, and the next morning the ship took on a quota of sick and wounded, 81 officers, 11 Nursing Sisters and 697 other ranks. Of these, 132 had either an arm or a leg missing. At noon on 17 March 1919, the *Czaritza*, a passenger steamer in peacetime, cast off and headed out into the Channel. Standing by the rail, Woodward felt the cool salt air on his face and looked back at the shore. It was not long before the coast of England disappeared into the sea mist and low cloud. He was finally on his way home. His mind went to his tunnellers – some still in camp and some in hospitals and convalescent homes, while others lay lifeless under the cold, wet French and Belgian earth. He was leaving them all behind.

In fair weather, the *Czaritza* made good time across a calm Bay of Biscay, but a following storm chased them into the Mediterranean as they headed east for Port Said in Egypt. Then it turned out there was a problem with the ship's heating system, which blew warm air throughout the ship: it was stuck on, and this would be a major problem once the ship headed south down the Suez Canal and into the tropics. After a delay of a week waiting in the Mediterranean, the *Czaritza* passengers were transferred to the *Dunluce Castle* and on 1 April they continued their journey south. The men were more than pleased with their new ship. It had been at the Gallipoli landing and was fitted out as a hospital transport vessel with a gymnasium, complete with a range of exercise equipment for wounded men. As on other returning transports, the men were well fed on good-quality food, had a range of amusements and sporting events, musical evenings and in some cases a special edition paper.

South through the Suez Canal and the Red Sea, they entered the Gulf of Aden. In Colombo, the men were granted shore leave. After racing around in a rickshaw for the first time, Woodward had lunch at the famous and luxurious Galle Face Hotel on the waterfront before journeying to Mount Lavinia, a popular Colombo seaside resort, which he found to be seriously inferior to many beaches he knew so well in Sydney.

Leaving Colombo on 23 April, the *Dunluce Castle* crossed the Equator and steamed south to Fremantle. While at dinner on 6 May, Oliver Woodward was told that at 9.30 that night, those onboard should pick up the faint flicker of the Rottnest Island Light. It was an excited group of men that crammed every vantage point to enjoy this first sight of the Australian coast. Right on time came the shrill call of the man on the lookout with, 'Light on port bow' and, sure enough, a little while later, there was the flashing navigation light far off on the horizon that was Rottnest

Island. 'I shall never forget the thrill occasioned by that light,' Woodward recalled in his diary. 'To us it meant home after the years of war, and I think the majority of us welcomed the darkness to hide the tears of joy which that flashing light brought to our eyes.'[7]

All that night the returning troops and nurses lined the rails watching for the slowly approaching coastline. As dawn broke, the ship anchored in the Gage Roads off Fremantle and the Western Australian contingent was taken off on a lighter. Everyone else was forced to stay onboard due to the concern of the influenza epidemic that was sweeping the world. On 7 May the *Dunluce Castle* set sail, arriving in Adelaide where the South Australians were disembarked before it again set sail, this time for Melbourne.

There, the men were given a thorough medical inspection and the Victorians and the Tasmanians were disembarked. The ship moved off from the New Pier and headed into the open sea and north to Sydney. Waking early, Woodward found the ship tracking up the beautiful south coast. The seas were unpleasantly rough, but nothing could dampen the excitement of the men who were about to arrrive home to friends and loved ones. One can imagine the joy they felt when the Macquarie Lighthouse was identified and the ship slipped into the calm, welcoming waters of Sydney Harbour. Just as when they had left, the lights of the South Head Signal Station welcomed them home as they anchored off Clifton Gardens for the night.

That night, no one aboard could sleep with the anticipation of their homecoming the next morning. At 9.30 am on Sunday 18 May 1919, the *Dunluce Castle* tied up alongside No. 1 wharf in Woolloomooloo Bay, three years and 87 days since Oliver Woodward had departed. After a round of farewells to his shipboard friends, he stepped ashore and was taken by one of a fleet of private cars to the Anzac Buffet in the Domain. 'We were set down at the entrance of a lane which led to the Buffet, and as I walked along I caught sight of my Mother and sister and then I knew what joy could really mean,' he wrote.[8]

After completing his discharge from the AIF, he was united with his mother and sister. They returned by rail to Tenterfield for a massive family reunion. As he stepped from the train, he was greeted at the station by his father and brother, and soon they were back home, 'in an unbroken family circle', he wrote. 'In all this great joy we did not forget the homes in Australia which would not experience this full reunion.' Eighty men from the three tunnelling companies would not return. Of 49 officers, six were killed and ten were wounded. Of 166 NCOs, 14 were killed and 39 were wounded. And of the 1980 sappers, 60 were killed and 241 wounded.

In decorations the company was awarded one Distinguished Service Order, 12 Military Crosses including the three awarded to Oliver Woodward, four Distinguished Conduct Medals, 29 Military Medals, nine Mention in Dispatches, 16 Mention in Orders and 13 Special Mention of the Company as a whole in Dispatches and Orders of the Day.

In all it was an impressive tally for a group of men who comprised only about one per cent of the AIF. But for all this few knew about the tunnellers' work and they received very little public recognition or praise, with the exception of a monument at Hill 60 to the 1st Australian Tunnelling Company.

Today, with the screening of the film *Beneath Hill 60*, the heroic effort of these men is again in the nation's consciousness – and deservedly so. Their terrifying work brought a quicker end to the war and saved Allied lives. We have much to be thankful for.

EPILOGUE

Oliver Woodward spent some time with his family at Tenterfield before going to work again at the Mount Morgan Gold Mining Company.

At the end of 1919, he travelled to Cairns to have Christmas with the Waddell family. 'It was a joy to meet Marjorie again after four years,' he wrote. 'When we parted she was a teenager with pig-tails; now she was a young woman, a transformation which seemed to lessen the gap in ages.' The two had been writing letters to each other regularly during the war years, and this, Woodward believed, ensured their reunion 'was free from embarrassment'. 'On Christmas eve,' wrote Woodward in his diary, 'Marjorie and I went for a long walk on [a] wonderful stretch of beach and there we became engaged.' In early January 1920, Woodward sailed to Sydney, and Marjorie farewelled him from the town wharf. 'I said a fond farewell to the girl I had loved for years.'[1]

He had secured a job at Port Pirie, in South Australia, as the General Metallurgist with Broken Hill Associated Smelters, where Sir Colin Fraser, the ex-geologist from the Laloki mine in Papua, was the General Manager. In late August, he travelled from there to Brisbane and met Marjorie and her family, who had taken the ship from Cairns. On the 3rd, the couple were married in a quiet, small ceremony at St John's Anglican Cathedral in Brisbane. They had their honeymoon in Blackheath, in the Blue Mountains, west of Sydney.

They made their home in Port Pirie, where Woodward had been recently promoted to Plant Superintendent. In September 1921 the couple's first child, Barbara, was born. She was followed in August 1924 by Oliver Gordon, who was born at home. Their third child, Colin Holmes, was born in October 1927. Marjorie dedicated her life to their children. 'No mother could have given more and loving attention to her children than she did for our three,' wrote Woodward.[2]

Woodward was appointed a member of the Council of the Port Pirie Technical College, but when Labor came to power in the state elections he was quickly removed from this position. 'This action I took as a warning that it would be unwise for one in my position to take too permanent a part in the civil life of this industrial town,' he wrote.[3]

In 1926 he was promoted to General Superintendent, a position he held for nine years. During this time he undertook extensive re-building at the facility and oversaw major improvements to the conditions of the working men.

In 1934 the family moved to Broken Hill, where Woodward became the General Manager of North Broken Hill Limited. For the following 13 years, he oversaw and managed a major re-building and modernisation program and also the re-opening of the British Broken Hill mine that had been closed in 1930, the early days of the Depression.

In 1940 Woodward became the President of the Australian Institute of Mining and Metallurgy and in 1944 was elected to the

Board of North Broken Hill, where he stayed until his retirement in October 1947 aged 62 years. During this time he was also a Director of Broken Hill Smelters and Electrolytic Refining and Smelting Company and from 1952 to 1954, the President of the Australian Mining and Metals Association.

In 1956 he was appointed CMG (a Companion of the Order of St Michael and St George) and died in Hobart, where he had retired, on 24 August 1966, at the age of 80. He was survived by Marjorie, who passed away on 30 July 1978, and his three children – Barbara, Oliver and Gordon.

ACKNOWLEDGEMENTS

My first thanks go to Ross Thomas from Townsville, who has devoted the last twenty years to researching and publicising the story of the Australian Tunnelling Companies of the First World War. It was Ross's work that initially led to the production of the feature film *Beneath Hill 60* and to the construction of a memorial to the tunnellers to be unveiled in the near future in Townsville, Queensland.

My thanks to Bill Leimbach, the producer of *Beneath Hill 60*, who along with David Roach, the co-producer and screenwriter, invited me to write the book. Their support and encouragement of the writing and editing period has been much appreciated.

Thanks go to Barbara Woodward and her family for making available her father's diary and for providing family photographs for the book. This diary is probably the greatest personal legacy and individual story of any Allied tunneller of the Great War and is quoted in most books on the subject and in the Official History.

I would like to especially acknowledge two books that were valuable references during the research and writing period, and to which I have often referred. First, the definitive history of the tunnellers by Captain W. Grant Grieve and Bernard Newman, titled *Tunnellers: The Story of the Tunnelling Companies, Royal Engineers, During the World War,* and a book by Alexander Barrie, *War Underground: The Tunnellers of the Great War.*

My thanks to Margaret Lewis and the staff of the Research Centre at the Australian War Memorial in Canberra and to the past head, Mal Booth, for advice, research information and making available files and reference materials. Also to the National Archives of Australia for making available online the personal files of our troops from the Great War, and to the Australian War Memorial for being the wonderful repository it is for Australian military history and tradition.

My thanks to David Heidtman for his legal advice and friendship, to Virginia Gordon for the loan of her volumes of Charles Bean's *Official History* and to Robyn Batchelor for proofreading my early drafts.

To my publisher, Meredith Curnow, and editors Vanessa Mickan and Sophie Ambrose, and the staff at Random House Australia for their ongoing support, advice and encouragement and for the amazing marketing initiative they put behind all their books and all of their writers.

And lastly my thanks to my wife, Heather, and my boys and all their family for their love and support over the difficult and isolating period of writing. Your assistance and understanding is greatly appreciated.

GLOSSARY

AIF	Australian Imperial Force
BEF	British Expeditionary Force
Block	Defended barricade in a trench
Camouflet	An underground explosive charge used to damage the enemy's underground workings but not break through to the surface. The word is derived from the Latin *calamo flatus* – a blast through a reed or a pipe
Clay kicking	A method of digging through clay where a miner lies on a cross-bar and uses a special curved blade shovel with his feet to work the face
CO	Commanding Officer
Colour patches	Distinctive badges worn by battalions and AIF units on each shoulder to differentiate them by colour, division and battalion

CSM	Company Sergeant Major
Detonator	A very small ignition charge that sets off the main explosive charge
Duckboard	Wooden decking
Face	The end or furthest extent of the mine tunnel where the work is being undertaken
Fire step	Step in the side of the trench to raise a man to a firing position
Funk hole	Hole in the side of a trench for sleeping and protection
Fuse	A device containing a combustible material for igniting a detonator, or an electric wire that carries a charge that ignites a detonator
Gallery	A tunnel or a large area at the end of a tunnel where the explosive charges are packed
Geophone	A French invention for listening underground to enemy activity
GHQ	General Headquarters
Head cover	The distance between the tunnel and the surface of the ground
HQ	Headquarters
Listening post	A position established to listen for the enemy's movements and give warning of an enemy attack
Mills grenade	British-issue hand grenade
Moles	English term for the 'clay kickers' – the miners who specialised in working in clay

NCO	Non Commissioned Officer
No-man's-land	The dangerous land between two opposing lines
Parapet	Front edge of a trench, a rampart to protect men in the trench above the level of the surrounding ground
Pioneers	Army engineers
Platoon	Army unit of 38 men under a lieutenant and sergeant
Proto set	A self-contained breathing apparatus for mine-rescue work
RE	Royal Engineers
Respirator	Gas regulator
Reveille	Dawn wake-up bugle call
Route march	Hard marching between two points
Russian saps	Underground tunnels close to the surface and towards the enemy that during an attack can protect infantry and be broken open to form a ready-made trench
Salient	A prominent or projecting part of the line often protruding out from the main frontline
Sam Browne	A leather belt worn as part of an officer's uniform
Sap	A trench dug towards the enemy from which trenches then radiate out on each side
Section	Ten men usually under the command of a corporal. One-third of a platoon
Shaft	A vertical tunnel down into a mine

Slurry	Mud, sand or other soils saturated with water and in a semi-liquid state
SRD	Strong rum, dilute. Sometimes also said to stand for Service Ration Depot
Tamping	Packing placed in a tunnel to force the explosive upwards and to prevent the force of the explosion blowing back along the gallery
Tapes	Cotton tapes laid down to designate the starting line for an attack
Tunnel	A horizontal gallery in a mine
Wire	Barbed wire
Wiring party	Group of men who put up barbed wire

WEIGHTS AND MEASURES

Linear Measure

1 inch	= 1000 mils	= 2.54 centimetres
12 inches	= 1 foot	= 0.3048 metre
3 feet	= 1 yard	= 0.9144 metre
220 yards or 660 feet	= 1 furlong	= 201.168 metres

Weights

	1 ounce	= 28.3495 grams
	1 pound	= 453.59 grams
14 pounds	= 1 stone	= 6.35 kilograms

References

Chapter 1: Mobility to Stalemate

1 Woodward, Oliver, Unpublished Diary, Vol. II, p. 6
2 Ibid., p. 8
3 Australian Dictionary of Biography
4 Woodward, Vol. II, p. 9

Chapter 2: The War Goes Underground

1 Barrie, Alexander, *War Underground*, pp. 110–111
2 Grieve, W. G. & Bernard Newman, *Tunnellers*, pp. 28–29
3 Barrie, *War Underground*, p. 26
4 Ibid., p. 27
5 Ibid., p. 28
6 Ibid.
7 Ibid., pp. 30–31
8 Grieve & Newman, *Tunnellers*, p. 35

Chapter 3: In the Darkness and Mud

1 Barrie, *War Underground*, pp. 32–33
2 Ibid., p. 33
3 Grieve & Newman, *Tunnellers*, p. 39

4   Ibid., *Tunnellers*, pp. 39–40

5   Ibid., *Tunnellers*, p. 44

6   Bean, C. E. W., *Official History*, Vol. IV, pp. 949–950

Chapter 4: The Experience of Gallipoli

1   Bean, C. E. W., *Official History*, Vol. II, pp. 199–200

2   White, Capt. T. A., *History of the Thirteenth Battalion*, p. 39

3   Ibid.

4   Bean, *Official History*, Vol. II, p. 233

5   Ibid., p. 279

6   Ibid., p. 203

7   Ibid., p. 233

8   Ibid., p. 280

9   Ibid.

10 Ibid., p. 327

11 Ibid., pp. 280–281

12 McNicoll, Ronald, *The Royal Australian Engineers*, p. 44

13 Bean, *Official History*, Vol. I, p. 60

14 McNicoll, *The Royal Australian Engineers*, pp. 56–57

Chapter 5: White Feathers and the Call to Arms

1   Woodward, Vol. I, p. 3

2   Ibid.

3   Woodward, Vol. II, pp. 9–10

4   Ibid., pp. 10–11

5   Ibid., p. 11

6   Ibid.

7   Ibid., p. 12

8   Ibid., p. 15

9   Ibid.

10 Woodward, Vol. V, p. 1

11 Ibid.
12 Ibid., p. 2
13 Ibid.
14 Ibid., p. 3
15 Ibid., p. 5
16 Woodward, Vol. II, p. 13
17 Ibid., pp. 12–13
18 Ibid., p. 15
19 Ibid., p. 14
20 Ibid.
21 Ibid.

Chapter 6: The Earthquake Idea

1 McNicoll, *The Royal Australian Engineers*, p. 50
2 Grieve & Newman, *Tunnellers*, p. 58
3 Ibid.
4 Barrie, *War Underground*, p. 72
5 Ibid., p. 73
6 Ibid., p. 74
7 Grieve & Newman, *Tunnellers*, p. 61

Chapter 7: Along the River Somme

1 Barrie, *War Underground*, pp. 101–102
2 Ibid., pp. 103–105
3 Ibid., p. 109
4 Grieve & Newman, *Tunnellers*, pp. 97–98
5 Barrie, *War Underground*, pp. 109–111
6 Ibid., pp. 126–131

Chapter 8: Misspent Energy and Wasted Effort

1 Grieve & Newman, *Tunnellers*, p. 73
2 Barrie, *War Underground*, p. 138
3 Ibid.
4 Ibid., p. 82
5 Grieve & Newman, *Tunnellers*, p. 98

Chapter 9: Off to the Western Front

1 Woodward, Vol. II, p. 17
2 Ibid., p. 18
3 Ibid.
4 Ibid., p. 19
5 Ibid.
6 Ibid., p. 22
7 Ibid., p. 23
8 Ibid., p. 25
9 Ibid., p. 26
10 Ibid.
11 Ibid., p. 27

Chapter 10: Back Near Messines

1 Bridgland & Morgan, *Tunnel-Master and Arsonist*, p. 198
2 Grieve & Newman, *Tunnellers*, p. 210
3 Ibid., p. 209
4 Ibid., p. 215
5 Ibid., p. 222
6 Barrie, *War Underground*, p. 218
7 Grieve & Newman, *Tunnellers*, p. 223
8 Ibid., p. 225
9 Ibid., p. 228
10 Ibid., p. 236

Chapter 11: Just out from Armentières

1 McNicoll, *The Royal Engineers*, p. 63
2 Woodward, Vol. II, p. 28
3 Ibid., p. 30
4 Ibid.
5 Ibid., p. 31
6 Ibid.
7 Ibid., p. 30
8 Ibid., p. 31
9 Ibid., p. 33
10 Ibid., p. 34
11 Bean, *Official History*, Vol. III, p. 217
12 Grieve & Newman, *Tunnellers*, pp. 110–111
13 Woodward, Vol. II, p. 37

Chapter 12: The Red House

1 Woodward, Vol. II, p. 39
2 Ibid.
3 Ibid., p. 40
4 Ibid., p. 41
5 Ibid., p. 42

Chapter 13: The First Big One

1 Grieve & Newman, *Tunnellers*, pp. 116–117
2 Ibid., p. 117
3 Ibid., p. 118
4 Ibid., p. 120
5 Ibid., p. 123
6 Ibid., pp. 126–127
7 Ibid., p. 128
8 Ibid., p. 129

Chapter 14: Sojourn at Ploegsteert Wood

1 Woodward, Vol. II, p. 45

2 Ibid., p. 46

3 Ibid.

4 Ibid.

5 Ibid.

6 Ibid.

7 Ibid.

8 Ibid.

9 Australian Dictionary of Biography

10 Barrie, *War Underground*, pp. 240–41

11 Bean, *Official History*, Vol. IV, p. 951

12 Ibid.

13 Woodward, Vol. II, p. 48

14 Ibid.

15 Ibid.

16 Ibid.

Chapter 15: The Move to Hill 60

1 Woodward, Vol. II, pp. 58–59

2 Ibid., p. 66

3 Ibid.

4 Ibid., p. 68

5 Ibid.

6 Woodward, Vol. III, p. 67

7 HQ Report 1 ATC, Ypres Salient, 1 December 1916

8 Ibid., 6 December 1916

9 Ibid., 8 December 1916

10 Ibid., 15 December

11 Ibid., December–January 1916/17

12 Woodward, Vol. II, p. 69

Chapter 16: A Month Today

1 Strachan, Hew, *The First World War*, p. 237
2 Woodward, Vol. II, p. 73
3 Bean, *Official History*, Vol. IV, p. 958
4 Woodward, Vol. II, p. 75
5 National Archives of Australia personal file
6 Woodward, Vol. II, p. 78
7 Ibid.
8 Ibid., p. 77
9 Ibid.
10 Bean, *Official History*, Vol. IV, p. 958
11 Ibid., pp. 958–959
12 Report by Oberstleutnant Otto Füsslein, entitled *Miners in Flanders*, in Heinrici, Paul

Chapter 17: The Days Before

1 Barrie, *War Underground*, p. 253
2 Grieve & Newman, *Tunnellers*, p. 229
3 Barrie, *War Underground*, p. 254
4 Woodward, Vol. II, p. 80
5 Quoted in Robins, Simon, *British Generalship on the Western Front*, p. 57
6 Cave, Nigel, *Hill 60: Ypres*, p. 101
7 Royal Engineers' Institute, *Military Mining*, pp. 40–41
8 Oberstleutnant Otto Füsslein, *Miners in Flanders*, in Heinrici, Paul, p. 13
9 Woodward, Vol. II, p. 81

Chapter 18: After the Earthquake

1 Bean, *Official History*, Vol. IV, p. 595
2 Ibid.
3 Ibid., p. 679

4 Royal Engineers' Institute, *Military Mining*, p. 42

5 Ibid.

6 Ibid., pp. 42–43

7 Bean, *Official History*, Vol. IV, p. 954

8 Quoted from a report by Oberstleutnant Otto Füsslein and taken from Heinrici, Paul, *Das Ehrenbuch der Deutschen Pioniere*, 1931, p. 541

9 Woodward, Vol. II, p. 82

Chapter 19: A New Role and a New Danger

1 Woodward, Vol. II, p. 90

2 Ibid.

3 Bean, *Official History*, Vol. IV, p. 761

4 Woodward, Vol. II, p. 94

5 Reid, Richard, *Beaucoup Australiens Ici*, p. 24

6 Woodward, Vol. II, p. 96

7 Ibid., p. 97

8 Ibid.

9 Ibid.

10 Ibid., p. 98

11 Ibid., p. 100

12 Ibid., p. 101

13 Ibid., pp. 101–102

14 Ibid., p. 102

15 Ibid.

16 Ibid., p. 103

17 Ibid.

18 Ibid., p. 106

Chapter 20: The Last Battle

1 Woodward, Vol. II, p. 83

2 Ibid., p. 108

3   Ibid.
4   Ibid., p. 111
5   Ibid., p. 113
6   Supplement to the *London Gazette* (31089), 9 December 1919, p. 15295
7   AWM Encyclopaedia Statistics – Military
8   Woodward, Vol. II, pp. 114–115
9   Ibid.

Chapter 21: Going Home

1   Woodward, Vol. II, pp. 117–118
2   Ibid., p. 117
3   Ibid., Vol. II, p. 119
4   Ibid., Vol. II, pp. 118–119
5   Ibid., Vol. II, p. 125
6   Ibid., p. 126
7   Ibid., p. 135
8   Ibid., p. 136

Epilogue

1   Woodward, Vol. V, pp. 7–8
2   Woodward, Vol. I, p. 34
3   Ibid.

BIBLIOGRAPHY

Addison, Col. G. H., 'The German Engineer and Pioneer Corps – Part I and II', *Royal Engineers Journal*, June and September 1930

Australian Dictionary of Biography (online resource)

Australian War Memorial Encyclopaedia Statistics (online resource)

Barrie, Alexander, *War Underground: The Tunnellers of the Great War*, Spellmount Ltd, Staplehurst, 2000

Barton, Peter, *The Battlefields of the First World War*, Random House Australia, Sydney, 2005

Barton, Peter, Peter Doyle & Johan Vandewalle, *Beneath Flanders Fields: The Tunnellers' War 1914–1918*, Spellmount Ltd, Staplehurst, 2004

Bean, C. E. W., *Official History of Australia in the War of 1914–1918*, Ninth Edition, Angus and Robertson, Sydney, 1940

Bridgland, Tony & Anne Morgan, *Tunnel-Master and Arsonist of the Great War: The Norton-Griffiths Story*, Leo Cooper, Barnsley, 2003

Cave, Nigel, *Hill 60: Ypres*, Battleground Europe series, Pen and Sword, Barnsley, 2000

Clifford, Col. M. D., 'Early History of Sapper Tunnelling', *The Royal Engineers' Journal* (given first as a speech by Col. Clifford on 16 October 1975)

Controller of Mines' General Report, *Mining Operations Messines Offensive*, Second Army BEF, June 1917

Dennis, Peter & Geoffrey Grey, *1917: Tactics, Training and Technology*, Chief of Army History Conference, 2007

East, Sir Ronald, *The Gallipoli Diary of Sergeant Lawrence of the Australian Engineers*, Melbourne University Press, Melbourne, 1981

Extracts from the Old Mining Regulations, Issued by the General of Pioneers, Army Headquarters, Laon, April 1915 (translation of a German pamphlet captured at Fricourt, July 1916)

Gammage, Bill, *The Broken Years*, Penguin Books Australia, Melbourne, 1975

German Mining Officer's Diary, captured Fricourt, July 1916

Grieve, W. G. & Bernard Newman, *Tunnellers: The Story of the Tunnelling Companies, Royal Engineers During the World War*, Herbert Jenkins Ltd, London, 1936

Guilliatt, Richard & Peter Hohnen, *The Wolf*, William Heinemann, Sydney, 2009

Hankey, Lord, *The Supreme Command 1914–1918*, Vol. I, Allen & Unwin, London, 1961

Harvey, Major-General R. N., *Mining in France 1914–1918*, Pro Wo 106/387

Heinrici, Paul, *Das Ehrenbuch der Deutschen Pioniere*, Berlin, 1931

Higgins, Paul, *The Australian Mining Corps in the Great War*, University College, University of New South Wales ADFA, Canberra, 1995

Hosken, Graeme, *Digging for Diggers: A Guide to Researching Australians of the Great War 1914–1918,* ANZAC Day Commemoration Committee, Queensland, Aspley, 2002

Inspector of Mines, *Mining Operations Messines Offensive,* June 1917

Ivor Evans, D., *Mining Warfare,* South Wales Institute of Engineers, 1920

Johnson, J. H., *Stalemate: The Real Story of Trench Warfare,* Rigel Publications, London, 1995

MacLeod, Roy, 'Phantom Soldiers: Australian Tunnellers on the Western Front 1916–18' *Journal of the Australian War Memorial,* 1988, Vol. 13

McMullin, Ross, *Pompey Elliot,* Scribe Publications, Melbourne, 2002

McNicoll, Ronald, *The Royal Australian Engineers 1902 to 1919,* Corps Committee of the Royal Australian Engineers, Canberra, 1979

Neill, J. C., *The New Zealand Tunnelling Company 1915–1919,* Whitcombe and Tombs Limited, Auckland, 1922

Oldham, Peter, *Messines Ridge: Ypres,* Battleground Europe series, Pen and Sword, Barnsley, 2003

Prior, Robin & Trevor Wilson, *The First World War,* Cassell, London, 1999

Reid, Richard, *Beaucoup Australiens Ici: The Australian Corps in France 1918,* Department of Veterans Affairs, Canberra, 1998

—— *Ypres 1917: Australians on the Western Front,* Department of Veterans Affairs, Canberra, 2007

Robins, Simon, *British Generalship on the Western Front 1914–18: Defeat into Victory,* Routledge, 2005

Roll of Honour: Register of the Tunnelling Companies 1915–1918

Royal Engineers' Institute, *Military Mining: The Work of the Royal Engineers in the European War 1914–19*, W. & J. MacKay Limited, Chatham, 1922

Stevenson, Lt-Col A., *Report of the Second Army Tunnelling Companies*, Central Headquarters, 21 June 1917

Strachan, Hew, *The First World War*, Simon and Schuster, London, 2003

Varley, P. M., *British Tunnelling Machines in the First World War*, Science Museum, London, 1993

White, Capt. T. A., *History of the Thirteenth Battalion AIF*, Sydney: 13th Battalion, A.I.F. Committee, 1924

Williams, John F., *Corporal Hitler and the Great War 1914–1918*, Frank Cass, London and New York, 2005

Woodward, Oliver, Unpublished Diaries, Vols. I–V

Somme Mud

The experiences of an infantryman in France, 1916–1919
E. P. F. Lynch
Edited by Will Davies

'We live in a world of Somme mud. We sleep in it, work in it, fight in it, wade in it and many of us die in it . . .'

Private Edward Lynch enlisted in the army aged just eighteen. As his ship set sail for France, the band played and the crowd proudly waved off their young men. Men who had no real notion of the reality of the trenches of the Somme; of the pale-faced, traumatised soldiers they would encounter there; of the mud and blood and the innumerable contradictions of war.

Upon his return from France in 1919, Private Lynch wrote about his experiences in twenty school exercise books, perhaps in the hope of coming to terms with all that he had witnessed there. Now published here for the first time, his story vividly captures the horror and magnitude of the war on the Western Front as experienced by the ordinary infantryman.

Told with dignity, candour and surprising wit, *Somme Mud* is a testament to the power of the human spirit – for out of the mud that threatened to suck out a man's very soul rose this remarkable true story of humanity and friendship.

'This is a warrior's tale . . . a great read and a moving eye-witness account of a living hell from which few emerged unscathed'
DAILY EXPRESS

9780553819137

In the Footsteps of Private Lynch

Will Davies

With a mighty roar the shell explodes spouting flame and phosphorous fumes everywhere. Mud is showered over everyone as pieces of shell fly over prone bodies. A man five feet ahead of me is sobbing – queer, panting gasping sobs. He bends his head towards his stomach just twice and is still. We've had our baptism of fire, seen our first man killed . . .'

Will Davies' discovery of the manuscript for *Somme Mud* revealed a lost treasure. Private Edward Lynch's personal account of the war in the trenches has since become a classic.

This companion volume is a fascinating contextual history of the war in France as experienced by an eighteen-year-old soldier and his comrades. In retracing the progress of Lynch and the 45th Battalion, the AIF – from the long route marches to flea-ridden billets, into the frontline against the enemy at such infamous places as Messines and Villers-Bretonneux, and on to the great push for victory after August 1918 – Will Davies shines unique light on life and death amidst the barbed wire and the mud, and the ebb and flow of the war on the Western Front.

And in revisiting these battlefields today and attempting to understand their significance, this book pays tribute to the young men who sacrificed so much over ninety years ago.

9780553824155

Gallipoli

L. A. Carlyon

'Because it was fought so close to his old home ground, Homer might have seen this war on the Gallipoli Peninsula as an epic. Brief by his standards, but essentially heroic. Shakespeare might have seen it as a tragedy with splendid bit-parts for buffoons and brigands and lots of graveyard scenes. Those thigh bones you occasionally see rearing out of the yellow earth of Gully ravine, snapped open so that they look like pumice, belong to a generation of young men who on this peninsula first lost their innocence and then their lives, and maybe something else as well . . .'

Gallipoli remains one of the most poignant battlefronts of the First World War and L. A. Carlyon's monumental account of that campaign has been rightfully acclaimed and a massive bestseller in Australia. Brilliantly told, supremely readable and deeply moving, Gallipoli brings this epic tragedy to life and stands as both a landmark chapter in the history of the war and a salutary reminder of all that is fine and all that is foolish in the human condition.

'Incisive, emotionally-charged and visceral . . . blends a real feel for the fighting soldier with a firm grasp of the strangely beautiful countryside which saw such a bewildering mix of tragedy, missed opportunity and wasted heroism. A hard-hitting and heart-breaking book'
RICHARD HOLMES

'Superb . . . Carlyon's writing is so vivid that you almost imagine yourself present. A stunning achievement'
SAUL DAVID, *DAILY TELEGRAPH*

'Carlyon is a gifted writer . . . his book deserves to take its place alongside other classic accounts of Gallipoli. He conveys the beauty of the place and its ugliness 90 years ago'
JOHN KEEGAN, *DAILY TELEGRAPH*

'A brilliantly managed narrative and remarkably even-handed . . . a superb account'
TREVOR ROYLE, *GLASGOW HERALD*

9780553815061

The Wolf

Richard Guilliatt & Peter Hohnen

On 30 November 1916, an apparently ordinary freighter left harbour in Kiel, Germany. She would not touch land again for another fifteen months. It was the beginning of a voyage which was to take her around the world leaving a trail of devastation in her wake.

She had started life as a merchant ship, but by the time the German navy had finished with her, the *Wolf* was the most technologically advanced fighting boat of her generation. And yet through clever camouflage, she could pass for a common-or-garden steamship.

It was to be one of the most daring clandestine naval missions of modern times. She traversed three of the world's major oceans, destroyed more than thirty Allied vessels and captured over four hundred men, women and children, surviving on fuel and food plundered from captured ships. She sailed back unharmed fifteen months later.

'An extraordinary work of storytelling and scholarship. This is history brought vividly to life'
DOUG STANTON

'One of the strangest, and strangely thrilling, war-at-sea adventures I have ever read'
EVAN THOMAS, *NEWSWEEK*

'A cracking tale of humanity in an otherwise inhuman war'
NAVY NEWS

9780552157056

Under the Wire

William Ash

Bill Ash is one of a rare breed. In 1940, he sacrificed his American citizenship to join in the fight against Hitler. He became a fighter pilot, came to Britain and flew Spitfires in combat . . .

In March 1942, Bill was shot down over France. He survived, evading capture for months before being betrayed to the Gestapo in Paris. Tortured and sentenced to death as a spy, he was saved from the firing squad by the Luftwaffe and sent to Stalag Luft III, the infamous 'Great Escape' prisoner of war camp. It was from there that Bill began his extraordinary three year-long 'tour' of occupied Europe. One of a handful of POWs to attempt more than a dozen break-outs – over the wire, tunnelling under it, cutting through it or simply strolling out of a camp's gates in disguise – he is one of the war's greatest escapers.

Hailed by *The Times* as 'a story of bravery in the face of brutality, of comradeship, of a never-say-die attitude', *Under the Wire* is both one man's remarkable memoir and a tribute to an entire generation.

'Well written and exciting…there are passages in this book that make the reader want to stand up and cheer'
CHARLES ROLLINGS, AUTHOR OF *WIRE AND WALLS*

'A remarkable story . . . brilliantly told'
TONY RENNELL, CO-AUTHOR OF *THE LAST ESCAPE*

'Thoughtful, deep and poignant . . . a testament to man's deep-seated yearning to be free'
ROBERT WILCOX, AUTHOR OF *SCREAM OF EAGLES*

9780553817119

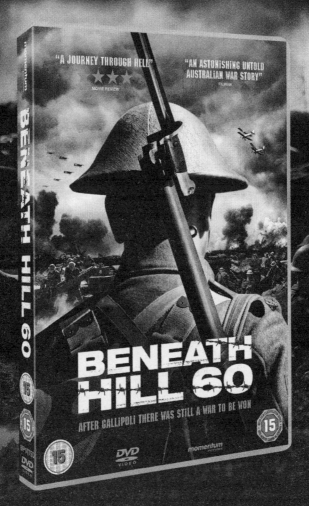